A First Course in Statistics

A Greer

**Formerly Senior Lecturer,
Gloucestershire College of Arts and Technology**

Stanley Thornes (Publishers) Ltd

First published 1980 by Stanley Thornes (Publishers) Ltd, Old Station Drive, Leckhampton, Cheltenham GL53 0DN, England

Reprinted 1980
Reprinted 1985 with minor corrections
Reprinted 1986
Reprinted 1988
Reprinted 1990

British Library Cataloguing in Publication Data

Greer, Alec
    A first course in statistics.
    1. Mathematical statistics
    I. Title
    519.5   QA276

    ISBN 0–85950–043–8

Typeset by European Engineering Services, Cheltenham and printed in Gt. Britain at The Bath Press, Avon.

PREFACE

This book provides a course in Statistics which includes the generality of topics required for first examinations in schools and colleges.

No prior knowledge has been assumed and the emphasis is on a simple approach to the subject. A teacher and the class may, if desired, work through the book, chapter by chapter.

A large number of graded exercises has been provided together with answers where these are appropriate. The student can thus work from relatively simple problems to a situation in which confidence in dealing with more difficult problems is acquired.

In addition there are three sets of miscellaneous exercises placed at strategic intervals throughout the book. All the questions in these exercises are of the type found in the examinations for which this book caters — short answer questions, standard questions and multiple choice questions. Hence the student may obtain practice in answering the various types of questions now common in many examinations.

A list of important formulae and data has been included in the hope that this will help students in their revision work. Finally a mathematical appendix containing proofs of some of the formulae used has been placed at the end of the book for the use of students who intend to take this subject at A-level and beyond.

In conclusion I wish to thank the staff of the publishers for their help in securing syllabuses and past examination papers and for their encouragement during the time when the script was in active preparation.

A. Greer

Gloucester 1980

CONTENTS

1 CHARTS AND DIAGRAMS

1.1 INTRODUCTION

Statistics is the name given to the science of collecting and analysing numerical facts. The facts, or data, may be obtained in a variety of ways most of which are discussed in Chapter 13.

In almost all scientific and business publications, in newspapers and Government reports facts and data are presented by means of tables, charts and diagrams. In this chapter the types of charts and diagrams most commonly used to present data are discussed.

All statistical charts and diagrams should:

(i)   be as simple as possible   (iv)   have their axes clearly labelled

(ii)   be as clear as possible   (v)   have a title

(iii) be easy to understand   (vi)   show the source of the information.

PERSONNEL EMPLOYED

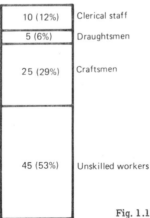

Fig. 1.1. The proportionate bar chart.

1.2 PROPORTIONATE BAR CHARTS

The proportionate bar chart (Fig. 1.1) relies on heights (or areas) to convey the proportions of a whole. The bar should be of the same width throughout its length (or height). The diagram has the following advantages:

(i)   It is accurate, quick and easy to construct.

(ii) It can show a large number of component parts without confusion.

It has the disadvantage that it may be difficult to make comparisons between the component parts visually. Although Fig. 1.1 shows the bar drawn vertically it can also be drawn horizontally if desired.

EXAMPLE 1. Table 1.1 shows the number of people employed on various types of work in a certain factory.

Table 1.1.

Type of personnel	Number employed
Unskilled workers	45
Craftsmen	25
Draughtsmen	5
Clerical staff	10
Total	85

Draw a proportionate bar chart to represent this information.

The easiest way is to draw the chart on graph paper. However, if plain paper is used the lengths of the component parts must be calculated and then drawn accurately using a rule (Fig. 1.1).

Suppose the total length of the bar is to be 12 cm. Then

$$85 \text{ people are represented by } 12 \text{ cm}$$

$$1 \text{ person is represented by } \frac{12}{85} \text{ cm}$$

$$45 \text{ people are represented by } \frac{12}{85} \times 45 = 6.35 \text{ cm}$$

$$25 \text{ people are represented by } \frac{12}{85} \times 25 = 3.53 \text{ cm}$$

$$10 \text{ people are represented by } \frac{12}{85} \times 10 = 1.41 \text{ cm}$$

$$5 \text{ people are represented by } \frac{12}{85} \times 5 = 0.71 \text{ cm}.$$

Alternatively, the proportions can be expressed as percentages which are calculated as shown on the opposite page.

Type of personnel	Percentage employed
Unskilled workers	$\dfrac{45}{85} \times 100 = 53\%$
Craftsmen	$\dfrac{25}{85} \times 100 = 29\%$
Draughtsmen	$\dfrac{5}{85} \times 100 = 6\%$
Clerical staff	$\dfrac{10}{85} \times 100 = 12\%$
Total	100%

1.3   SIMPLE BAR CHARTS

In these charts the information is represented by a series of bars all of the same width. The height, or length, of each bar represents the magnitude of the figures. The bars may be drawn vertically or horizontally as shown in Figs. 1.2 and 1.3 which present the data of Table 1.1.

These diagrams are also accurate, quick and easy to construct and they can show a large number of component parts without confusion. However, they do not readily display the total of all the component parts.

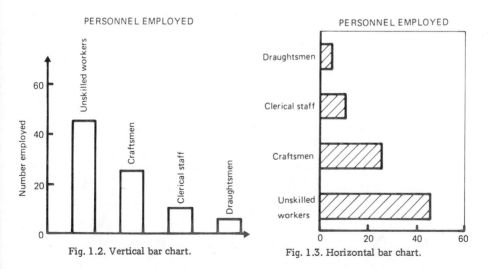

Fig. 1.2. Vertical bar chart.   Fig. 1.3. Horizontal bar chart.

1.4 CHRONOLOGICAL BAR CHARTS

These give a comparison of quantities over periods of time. They are essentially the same as vertical bar charts (Fig. 1.2) and are constructed in basically the same way as a graph.

EXAMPLE 2. The following data provide the annual figures for the number of colour television sets sold in Southern England for six successive years.

Year	Number of sets sold (thousands)
1	77.2
2	84.0
3	91.3
4	114.6
5	130.9
6	142.5

(Source: Post Office)

Draw a bar chart to represent this information.

When drawing a chronological bar chart, time is always plotted along the horizontal axis. The chart is shown in Fig. 1.4 and it shows clearly how the number of TV sets sold has increased over the period depicted.

NUMBER OF COLOURED TV SETS SOLD IN SOUTHERN ENGLAND
(Source: Post Office)

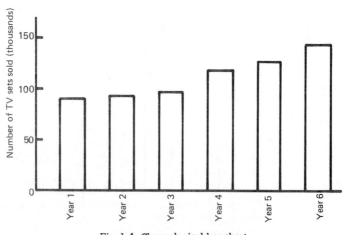

Fig. 1.4. Chronological bar chart.

1.5   COMPONENT BAR CHARTS

These are used when each total figure is made up of two or more component parts. The diagram is like a chronological bar chart except that the bars are divided into the component parts. The overall height of the bar and the heights of the divisions represent the actual figures.

EXAMPLE 3. A chemical company uses three raw materials, A, B and C in the manufacture of one of its products. The company's expenditure, in £ thousands, on these raw materials for four successive years are shown in the table below.

	Expenditure (£ thousands)			
	Year 1	Year 2	Year 3	Year 4
A	12	14	16	19
B	29	30	33	33
C	16	23	25	29
Totals	57	67	74	81

Construct a component bar chart to illustrate the yearly expenditures and their constituent parts.

Two types of component bar charts as shown in Fig. 1.5 may be drawn.

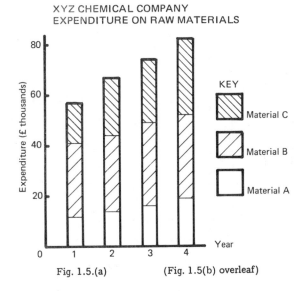

XYZ CHEMICAL COMPANY
EXPENDITURE ON RAW MATERIALS

Fig. 1.5.(a)   (Fig. 1.5(b) overleaf)

XYZ CHEMICAL COMPANY
EXPENDITURE ON RAW MATERIALS

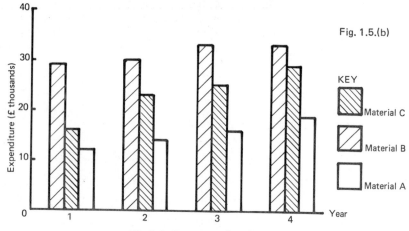

Fig. 1.5.(b)

KEY

Material C

Material B

Material A

Fig. 1.5. Component bar charts.

1.6   Sometimes a *percentage component bar chart* is required. The component parts are calculated as a percentage of the total as shown below for the data of Example 3. Note that the overall length of the bars are all the same because the total height of each bar represents 100% (Fig. 1.6).

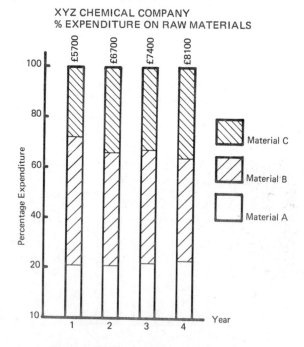

XYZ CHEMICAL COMPANY
% EXPENDITURE ON RAW MATERIALS

Material C

Material B

Material A

Fig. 1.6. Percentage component bar chart.

| | Percentage expenditure | | | |
	Year 1	Year 2	Year 3	Year 4
A	$\frac{12}{57} \times 100 = 21\%$	21	22	23
B	$\frac{29}{57} \times 100 = 51\%$	45	45	41
C	$\frac{16}{57} \times 100 = 28\%$	34	33	36
Totals	100%	100	100	100

Multiple and component bar charts should only be used when there are not more than four component parts. Any more tend to make the diagram too complicated and simplicity of presentation is then lost. When the data contain a large number of component parts a pie diagram (or a series of pie diagrams) should be used.

1.7 WHICH BAR CHART TO USE

(i)   Simple bar charts are used when changes in totals are required.

(ii)   The proportionate bar chart is used when there is only one total and the component parts are to be represented.

(iii)   Component bar charts are used when both the changes in totals and the component parts of each total are to be shown.

(iv)   Percentage component bar charts are used when only the relative changes in the component parts are required.

(v)   Multiple bar charts are used when the actual values of the component parts are required and where the overall totals are of no importance.

(vi)   Chronological bar charts are used when the data relates to a period of time. They show the changes in totals for the period depicted.

1.8 PIE CHARTS

A pie chart displays the proportions of a whole as sector angles (or sector areas), the circle as a whole representing the total of the component parts.

EXAMPLE 4. Represent the data of Table 1.1 in the form of a pie chart.

The first step is to calculate the sector angles. Remembering that a circle consists of $360°$ the sector angles are calculated as shown below:

Type of personnel	Sector angle (degrees)
Unskilled workers	$\frac{45}{85} \times 360 = 191$
Craftsmen	$\frac{25}{85} \times 360 = 106$
Draughtsmen	$\frac{5}{85} \times 360 = 21$
Clerical workers	$\frac{10}{85} \times 360 = 42$

Using a protractor the diagram may now be drawn (Fig. 1.7). If desired, percentages can be displayed on the diagram.

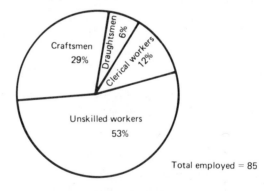

Fig. 1.7. A pie chart

1.9   Pie charts are very useful where proportions of a whole are to be represented. Up to about eight component parts can be accommodated, but above this number the effectiveness of the diagram decreases. Pie charts can be used to compare chronological data, but a separate pie chart is needed for each period.

EXAMPLE 5. The table on the following page shows the expenditure, in £, on various items purchased for wine and cheese evenings held in two successive years by a social club.

	Expenditure (£)	
	Year 1	Year 2
Wine	35	85
Cheese	18	33
Salads	9	12
Meat	22	35
Bread	2	3
Totals	86	168

Construct a pie chart for each year so that total costs may be compared.

When total costs are to be compared circles whose areas are proportional to the total costs are used. Now the area of a circle is $\pi \times$ radius^2. Therefore, if the radius of the circle representing the total costs for 1974 is made equal to 5 cm then, if R is the radius of the circle representing total costs for 1975

$$\frac{\pi \times R^2}{\pi \times 5^2} = \frac{168}{86}$$

$$R^2 = \frac{168}{86} \times 5^2$$

$$R = 5 \times \sqrt{\frac{168}{86}} = 7 \text{ cm (approx.)}$$

The component parts of each total are then used, as before, to give the various sectors of the two circles as shown in Fig. 1.8.

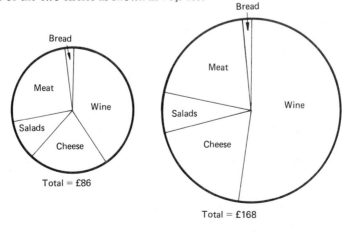

Fig. 1.8. Pie charts used to compare total costs.

If the total costs are not important then the two circles would be drawn with the same radii.

1.10 PICTOGRAMS AND IDEOGRAPHS

These are diagrams in the form of pictures which are used to present information to those who are unskilled in dealing with figures or to those who have only a limited interest in the subject.

Fig. 1.9 shows a typical pictogram (or picturegram). An almost limitless variety of symbols can be used to represent values. Each symbol in the pictogram represents a certain amount; in Fig. 1.9 each symbol represents 100 TV transmitters. One disadvantage of the pictogram is its lack of precision but this can be overcome by stating the values represented as shown in Fig. 1.9.

WORLD TELEVISION COVERAGE

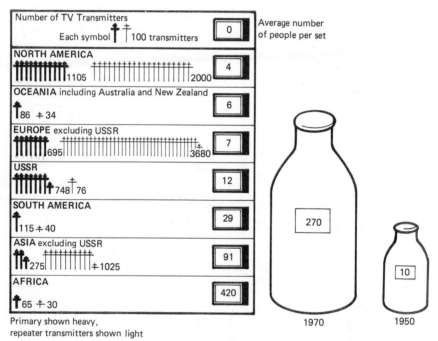

Primary shown heavy,
repeater transmitters shown light

(Courtesy: Barclays Bank Review).

Fig. 1.9. A pictogram

Fig. 1.10. Sales of milk in 1950 and 1970 (millions of litres).

A method not recommended is shown in Fig. 1.10. Comparison is difficult because the reader is not certain whether to compare heights, areas or volumes. However, if this method is used then the methods of Examples 6 and 7 should be used.

A First Course in Statistics

EXAMPLE 6. If a square of 3 cm side represents a population of 18 000, what population is represented by a square of 4 cm side?

Here, the quantities are represented by the areas of the squares. Since the area of a square = side².

3^2 cm² represents 18 000

1 cm² represents $\dfrac{18\,000}{9}$ = 2000

4^2 cm² represents 2000 × 16 = 32 000.

Hence a square of side 4 cm represents a population of 32 000.

EXAMPLE 7. A production of 1000 tonnes is represented by a cube of side 2 cm. Calculate the production represented by a cube of side 3 cm.

Here the quantities are represented by the volumes of the cubes. Since the volume of a cube = side³,

2^3 cm³ represents 1000 tonnes

1 cm³ represents $\dfrac{1000}{8}$ = 125 tonnes

3^3 cm³ represents 125 × 27 = 3375 tonnes.

Hence a cube of side 3 cm represents a production of 3375 tonnes.

1.11 CARTOGRAPHS

These are simply maps marked in some way so as to convey statistical data. The cartograph shown in Fig. 1.11 shows the variations of wool production in Europe and the Middle East.

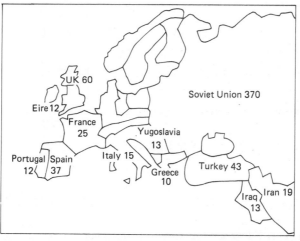

Fig. 1.11. Wool production in 1966 in Europe and the middle East
(figures are in millions of kilograms).

EXERCISE 1

1. In Fig. 1.12, find the values of x and y.

2. In Fig. 1.13, find the total expenditure and the expenditure on steel, brass and bronze.

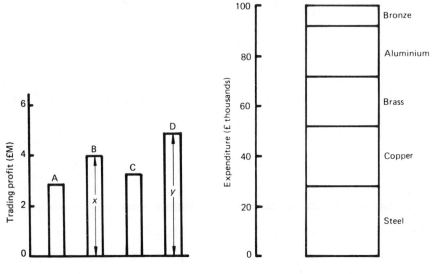

Fig. 1.12. Trading profits of companies.   Fig. 1.13.

3. Draw a proportionate bar chart for the following information which relates to expenditure per head on transport. In each case state the percentage expenditure and show them on your diagram.

Type of transport	Expenditure (£)
Private motoring	£1.10
Rail	£2.75
Other public transport	£3.15
Total	£7.00

4. The figures below show the way in which commuters in the South East region travelled to the London area in a certain year. The data were obtained by questioning 3000 commuters.

Type of transport	Numbers using
Private cars	1560
Bus and underground	840
British Rail	320
Other transport	280

Draw (a) a simple vertical bar chart

(b) a pie chart.

5. The information below gives details of the temperature ranges for forging various metals. Draw a horizontal bar chart to represent this information.

Metal	Carbon steel	Wrought iron	Brass	Copper
Forging range (°C)	770 – 1330	860 – 1340	600 – 800	500 – 1000

6. The figures below give the world population (in millions of people) from 1750 to 1950. Draw a chronological bar chart to represent this information.

Year	1750	1800	1850	1900	1950
Population	728	906	1171	1608	2504

(Source: UNESCO)

7. The figures below give the annual production of grain (in tonnes) for six successive years for the XYZ Farm Ltd. Draw a component bar chart to represent this information.

	Year 1	Year 2	Year 3	Year 4	Year 5	Year 6
Wheat	200	190	235	250	180	205
Barley	120	135	170	200	160	200
Oats	75	85	90	110	80	110
Total	395	410	495	560	420	515

8.   The data below give the areas of the various continents of the world.

Continent	Area (millions of km^2)
Africa	30.3
Asia	26.9
Europe	4.9
N. America	24.3
S. America	17.9
Oceania	22.8
USSR	20.5

(Source: United Nations)

Draw a pie chart to represent these data.

9.   A pie chart is drawn and the sector angles representing the information from which the chart was constructed are as follows:

Item	Sector angle
Food and drink	137°
Housing	61°
Transport	43°
Clothing	47°
Other	72°

The chart represents the way in which each pound of income was spent in 1968. Calculate the actual amounts spent per pound of income.

10.   Three pie charts are to be constructed which are to show the amounts spent by a firm on raw materials in the years 1977, 1978 and 1979. The total amounts spent were, respectively, £120 000, £180 000 and £220 000. If a circle whose radius is 5 cm is to be used to represent the total amount spent in 1977, find the radii of the circles which will be used to represent the expenditures in 1978 and 1979.

11.   The information on the opposite page shows the personal expenditure per head on transport in the years 1940, 1950 and 1960. Draw pie charts in such a way that the total expenditures per head can be compared.

Type of transport	Expenditure per head (£)		
	1940	1950	1960
Private motoring	2.64	3.63	11.13
Rail travel	1.08	1.98	2.52
Other public transport	2.28	5.39	7.35

12. A survey to establish the most liked sports of 300 boys gave the following results. Draw a pictogram to represent this information.

Sport	Athletics	Football	Cricket	Hockey	Swimming
Number	30	88	64	42	76

13. Fig. 1.14 is a pictogram depicting the population of the world from 1850 to 1950. From the diagram state, as near as you can, the population in 1900 and 1950.

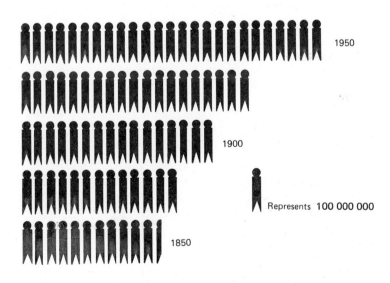

Fig. 1.14. World population.

14. The figures below show the production of bicycles of the W & P Cycle Co. Ltd. in a certain year. Draw a suitable pictogram to represent this information.

Month	Jan	Feb	March	April	May
Production (thousands)	32	27	18	24	30

2 GRAPHS

2.1 Graphs are an alternative form of pictorial representation. They are often to be preferred to the diagrams illustrated in Chapter 1.

2.2 AXES OF REFERENCE

To plot a graph we first take two lines at right angles to each other (Fig. 2.1). These lines are called the *axes of reference*. Their intersection, the point 0, is called the origin. The vertical axis is often called the y-axis and the horizontal axis is then called the x-axis. x and y are called variables.

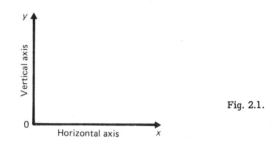

Fig. 2.1.

2.3 SCALES

The number of units represented by a unit length along an axis is called the scale on that axis. For instance 1 cm could represent 2 units. The scale is determined from the highest and lowest values to be plotted along an axis. It should be as large as possible but it must also be chosen so that it is easy to read.

The most useful scales are 1 unit, 2 units and 5 units of the variable to 1 large square on the graph paper. Some multiples of these, such as 10, 20, 50, 100 units etc. of the variable to 1 large graph paper square, are also suitable.

The scales chosen can be different on each of the axes.

2.4 COORDINATES

Coordinates are used to mark the points of a graph. In Fig. 2.2, values of x are to be plotted against values of y. The point P has been plotted so that $x = 8$ and $y = 10$. The values of 8 and 10 are said to be the rectangular coordinates of the point P. For brevity we say P is the point (8, 10).

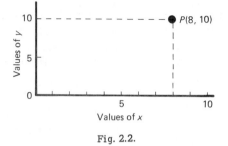

Fig. 2.2.

2.5   DRAWING A GRAPH

Every graph shows a relation between two sets of numbers. The table below gives the average diameter of ash trees of varying ages.

Age (years)	5	10	15	20	25	30	40	50	70
Diameter (cm)	7.6	9.3	12.2	16.2	21.4	27.7	43.8	64.5	119.7

To plot the graph we first draw the two axes of reference (Fig. 2.3). We then choose suitable scales to represent the age in years along the horizontal axis and the diameter along the vertical axis. Scales of 1 large square = 10 years (horizontally) and 1 large square = 20 cm (vertically) have been chosen. On plotting the graph we see that it is a smooth curve which passes through all of the plotted points.

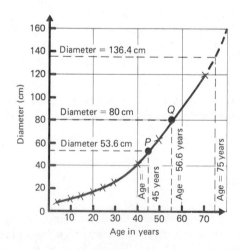

Fig. 2.3.

When a graph is either a smooth curve or a straight line we can use the graph to deduce corresponding values not given in the table of values. Thus to find the diameter of a tree which is 45 years old we first find 45 on the horizontal axis and from this point we draw a vertical line to meet the curve at P. From P we now draw a horizontal line to the vertical axis and read off the value. It is found to be 53.6. Hence a tree which is 45 years old will have a diameter of 53.6 cm.

Suppose now that we wish to know the age of a tree with a diameter of 80 cm. We find 80 cm on the vertical axis and from this point we draw a horizontal line to meet the curve at Q. From Q we draw a vertical line to the horizontal axis and read off the value. It is found to be 56.6. Hence a tree with a diameter of 80 cm is 56.6 years old.

Using a graph in this way to find values which are not given in the table is called *interpolation*. If we extend the curve so that it follows the general trend we can estimate values of the diameter and age which lie just beyond the range of the given values. Thus by extending the curve in Fig. 2.3 we can find the probable diameter of a tree which is 75 years old. This is found to be 136.4 cm.

Finding a value in this way is called *extrapolation*. An extrapolated value can usually be relied upon, but in certain cases it may contain a substantial amount of error. Extrapolated values must therefore be used with care. It must be clearly understood that interpolation can only be used if the graph is a smooth curve or a straight line. It is no good applying interpolation to the graph of Example 1.

EXAMPLE 1. The table below gives the temperature at 12.00 noon on seven successive days. Plot a graph to illustrate this information with the day horizontal.

Day June	1	2	3	4	5	6	7
Temp. ($^\circ$C)	16	20	16	18	22	15	16.5

As before we draw two axes at right-angles to each other, indicating the day on the horizontal axis. Since the temperatures range from 15° to 22°C we can make 15°C (say) our starting point on the vertical axis. This will allow us to use a larger scale on that axis which makes for greater accuracy in plotting the graph. Note that zero is shown by breaking the vertical axis.

On plotting the points (Fig. 2.4) we see that it is impossible to join the points by means of a smooth curve. The best we can do is to join the points by means of a series of straight lines. The graph then presents in pictorial form the variations in temperature and we can see at a glance that 1, 3 and 6 June were cool days whilst 2 and 5 June were warm days.

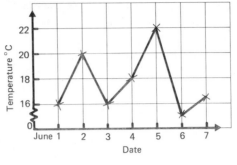

Fig. 2.4.

2.6 GRAPHS OF EQUATIONS

Consider the equation:

$$y = 2x + 5$$

We can give x any value we like and so calculate the corresponding value of y. Thus

when $x = 0, y = 2 \times 0 + 5 = 0 + 5 = 5$

when $x = 2, y = 2 \times 2 + 5 = 4 + 5 = 9$

when $x = 4, y = 2 \times 4 + 5 = 8 + 5 = 13$, and so on.

The value of y depends upon the value given to x and therefore, y is called the *dependent variable*. Since we can give x any value we please, x is called the *independent variable*. It is usual to mark the values of the independent variable along the horizontal axis and values of the dependent variable along the vertical axis.

EXAMPLE 2. Plot the graph of $y = 3x^2 - 2x + 5$ for values of x between 0 and 6.

A table may be drawn up as follows which gives values of y for the chosen values of x.

x	0	1	2	3	4	5	6
$3x^2$	0	3	12	27	48	75	108
$-2x$	0	-2	-4	-6	-8	-10	-12
$+5$	5	5	5	5	5	5	5
y	5	6	13	26	45	70	101

The graph is shown in Fig. 2.5, and it is a smooth curve. Graphs of mathematical equations always give either a smooth curve or a straight line.

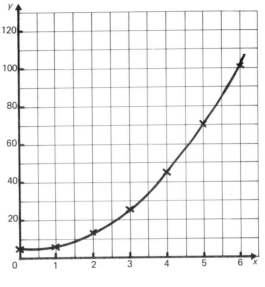

Fig. 2.5.

2.7   GRAPHS OF STATISTICAL DATA

In drawing graphs of statistical data the following points should be observed:

(i)   The title should be clearly stated on the graph.

(ii)   Both axes must be labelled with the name of the variable and its unit i.e. distance (metres), sales (thousands), etc.

(iii)   The independent variable must be placed on the horizontal axis. For instance, time is always the independent variable, and hence it is always taken along the horizontal axis.

(iv)   The source of the figures used must be stated. Sometimes the actual figures used are stated alongside the points on the graph (Fig. 2.6) but alternatively they can be placed as shown in Fig. 2.7.

(v)   The zero line on the vertical axis must always be shown but see Figs. 2.9 and 2.10.

(vi)   If percentage scales are used some indication of the values represented should be given.

(vii)   Care should be taken that the correct impression of the data is given on the graph (see Figs. 2.11 and 2.12).

Mathematical type graphs like the one shown in Fig. 2.3 can be used to obtain values between those given in the original information. Graphs of statistical data should not be used in this way except to obtain very approximate values. The main consideration of these graphs is that they should give a clear pictorial representation of the given information.

EXAMPLE 3. The following table shows the sales of a commodity during the years 1966 – 78.

Year	1966	1968	1970	1972	1974	1976	1978
Sales (£m)	1.2	1.8	2.2	3.4	5.6	8.1	9.8

Draw a graph of this information.

The graph is shown in Fig. 2.6. It shows that the amount of sales has been growing steadily during the period shown on the graph. The points on the graph have been joined by a series of straight lines because no information is given regarding the intermediate years.

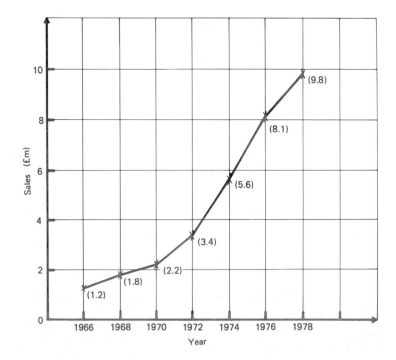

Fig. 2.6.

EXAMPLE 4. The information below relates to the imports of goods and services into the United Kingdom for the years 1973 − 78. Draw a graph of this information.

Year	1973	1974	1975	1976	1977	1978
Imports (£m)	18 965	27 410	29 028	36 793	42 390	45 777

(Source: Monthly Digest of Statistics)

The graph is drawn in Fig. 2.7. Note that the figures from which the graph was drawn are written at the top of the graph over the points to which they relate.

IMPORTS OF GOODS AND SERVICES INTO THE UK
(Source: Monthly Digest of Statistics)

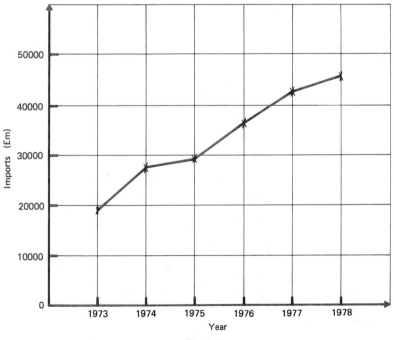

Fig. 2.7.

EXAMPLE 5. The figures below show the amount of grain grown by M & T Farms Ltd. in the Years 1 — 9. Represent these data *on a graph.*

Year	1	2	3	4
Grain grown (tonnes)	200	176	233	258

Year	5	6	7	8	9
Grain grown (tonnes)	243	197	200	265	238

Fig. 2.8 shows the graph plotted with the vertical scale starting from zero which results in a considerable amount of wasted graph paper between the actual graph and the horizontal axis.

Since the amount of grain grown varies from 176 to 265 tonnes we can choose a much larger vertical scale as shown in Fig. 2.9. Note, however, that the zero is shown at the bottom of the scale and that a definite break in the scale has been shown.

An alternative method of achieving the same objective is shown in Fig. 2.10 where a wavy line has been used to indicate the break in the vertical scale.

GRAIN GROWN BY M & T FARMS LTD.

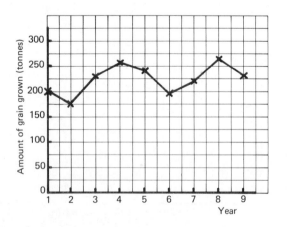

Fig. 2.8.

GRAIN GROWN BY M & T FARMS LTD.

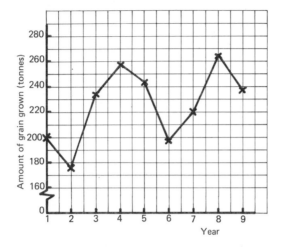

Fig. 2.9.

GRAIN GROWN BY M & T FARMS LTD.

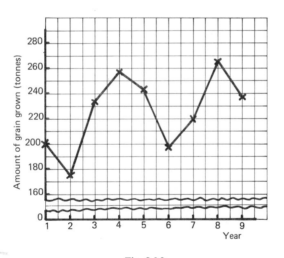

Fig. 2.10.

Unless care is taken over the choice of scales it is possible to give an entirely erroneous impression of the data being plotted.

EXAMPLE 6. The following data provide annual figures for the volume of sales of coloured television sets (Southern England) for a period of 5 years:

Year	Number of sets sold (Thousands)
1	77.2
2	84.0
3	91.3
4	114.6
5	130.9
6	142.5

Draw a suitable graph to represent these data.

SALES OF COLOURED TELEVISION SETS IN SOUTHERN ENGLAND

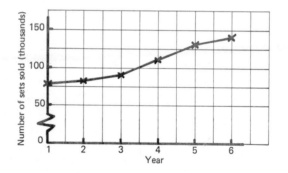

Fig. 2.11.

SALES OF COLOURED TELEVISION SETS IN SOUTHERN ENGLAND

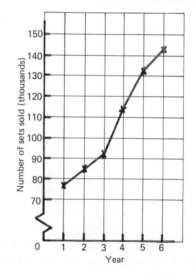

Fig. 2.12.

A First Course in Statistics

SALES OF COLOURED TELEVISION SETS IN SOUTHERN ENGLAND

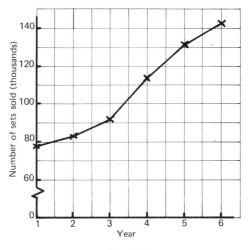

Fig. 2.13.

Three graphs of these data are shown in Figs. 2.11, 2.12 and 2.13. In Fig. 2.11 the vertical scale is too small and it appears that sales are increasing very slowly. In Fig. 2.12 the vertical scale has been made too large compared with the horizontal scale. As a result it appears that sales are increasing very rapidly. Fig. 2.13 appears to give the correct impression, i.e. that sales increased only slowly in the Years 1 − 3 but much more rapidly in the Years 3 − 6.

Several graphs may be drawn on the same piece of graph paper using the same axes and scales. Care must be taken, however, not to clutter the diagram with too many graphs or clarity will be lost. Example 7 shows the method.

EXAMPLE 7. The information below gives the quantities (in tonnes) of wheat, barley and oats grown on the XYZ Farms in Years 1 − 7. Draw three graphs of this information using the same axes and the same scales.

Year	1	2	3	4	5	6	7
Wheat	120	136	128	130	138	125	140
Barley	75	82	75	87	79	80	90
Oats	58	62	60	64	58	62	68

The three graphs are shown in Fig. 2.14 and it shows the variations in the production of the three types of grain grown on the farm.

GRAIN GROWN ON XYZ FARMS

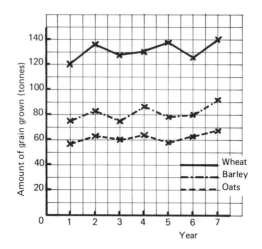

Fig. 2.14.

1. The table below gives the quantities of potatoes (in thousands of tonnes) grown in the United Kingdom during the years 1973 – 78. Draw a suitable graph to depict this information.

Year	1973	1974	1975	1976	1977	1978
Quantity	465	470	444	332	386	459

(Source: Monthly Digest of Statistics)

2. The figures below give the average total expenditure per week for the years 1973 – 78. Draw a graph of these data.

Year	1973	1974	1975	1976	1977	1978	1979
Expenditure (£ per week)	15	18	22	25	29	33	38

3. The following table gives details of sales, by value, of a trading firm for the year 1979. Draw a graph of these data.

Month	Jan	Feb	Mar	Apr	May	June
Sales (£thousands)	27	24	42	45	47	56

Month	July	Aug	Sept	Oct	Nov	Dec
Sales (£thousands)	61	52	68	73	80	72

4. The figures below give the advertising expenditure of a firm for 8 months in 1980. Draw a graph of these data.

Month	1	2	3	4	5	6	7	8
Expenditure (£thousands)	26	52	11	47	51	36	18	26

5. The data below show the sales made by a firm for the months Jan – Aug 1979. Draw a graph illustrating this information.

Month	1	2	3	4	5	6	7	8
Sales (hundreds)	102	142	135	115	156	171	127	110

6. The table below gives the total number of road casualties in the years 1971 – 77. Draw a graph of this information.

Year	1971	1972	1973	1974	1975	1976	1977
Casualties (thousands)	348	340	325	325	354	360	368

7. The table below shows the numbers of agricultural and non-agricultural workers in the United States in the years 1850 – 1950. Draw graphs on the same axes to illustrate this information.

Year	1850	1870	1890	1910	1930	1950
Agricultural workers (millions)	4.9	6.9	9.9	11.6	10.5	6.8
Non-agricultural workers (millions)	2.8	6.1	13.4	25.8	38.4	52.2

8. Details of advertising expenditure and sales by a trading company are shown in the table below. Draw two graphs on the same axes to illustrate this information.

Month	1	2	3	4	5	6	7	8
Advertising Expenditure (£thousands)	52	104	36	94	102	72	36	52
Sales (hundreds)	205	287	270	230	326	354	250	220

9. Draw graphs of the following equations:

    (a) $y = 5x + 2$ for values of x between 0 and 4.

    (b) $y = 8 - 2x$ for values of x between 0 and 4.

    (c) $y = 2x^2 + 3x + 1$ for values of x between 0 and 5.

    (d) $y = 5x^2 - 2x + 4$ for values of x between 0 and 6.

10. The table below gives the values of £1 invested at 8% compound interest for various periods of time:

Years	2	4	6	8	10	12	14
Value (£)	1.17	1.36	1.59	1.85	2.16	2.52	2.94

Plot a graph of this information drawing a smooth curve through the points with time on the horizontal axis. Use the graph to estimate:

(a) the value of £1 after 7 years

(b) the value of £1 after 15 years.

11. The table below gives corresponding values of x and y. Plot a graph and from it estimate:

(a) the value of y when x = 1.5

(b) the value of x when y = 30.

x	0	1	2	3	4	5
y	3	5	11	21	35	63

MISCELLANEOUS EXERCISE

EXERCISE 3

(All questions are of the type usually found in O-level, CSE and similiar examination papers.)

1. The table below shows the number of live births and marriages in Wales (in thousands) for the years 1973 − 78. Draw time series graphs to illustrate these data.

Year	1973	1974	1975	1976	1977	1978
Births	37.6	36.2	34.0	33.4	31.8	32.0
Marriages	22.3	21.3	20.6	19.5	19.7	20.4

(Source: Office of Population Censuses and Surveys)

2. The table below shows the crude oil production and the reserves available for future production of four leading oil exporting countries at the end of 1973. Calculate the further years of production at 1973 level for Iran, Venezuela and Kuwait. Give your values to the nearest year.

Country	1973 production (million barrels)	Oil reserves (million barrels)	Further years of production
Saudi Arabia	2800	140 700	50
Iran	2190	60 000	
Venezuela	1230	14 000	
Kuwait	1148	72 750	

Draw a bar diagram suitable for comparing the further years of production of these countries.

3. The table below shows the amount of energy consumed (in millions of tonnes of coal or coal equivalent) in the United Kingdom in 1978.

Coal	Petroleum	Gas	Nuclear electricity
120	139	64	13

Represent this information in a circular (pie) diagram of radius 5 cm.

4. The table below refers to two items of food production for the United Kingdom and France in a particular year.

	Amount (1000 tonnes)	
	Wheat	Fish
UK	3 400	1 100
France	14 500	750

Draw suitable pictograms to illustrate the data.

5. Draw a graph to show the points indicated in the table below which shows the values of x and y obtained from a statistical survey. Use a scale of 2 cm = 2 units on the horizontal (x) axis and 2 cm = 5 units on the vertical (y) axis. Join the points with a smooth curve.

Use your graph to find the value of y when x = 8.7 and the value of x when y = 30.

x	½	1	2	3	4	5	6	7	8	9	10	11	11½
y	1	2	5	13	24	35	40	35	24	13	5	2	1

6. Alvin Yardbrush, the pop singer, feels that not enough of the money paid by those who buy his latest LP record comes to him. He checks up, and finds that, of the £2.40 paid for each record:

    88 p goes to the shop
    24 p goes to the Government in tax
    16 p goes to the song writers
    87 p goes to the record company
    The rest goes to Alvin.

(a) Draw a pie chart in a circle of 5 cm radius to show how the money paid for the record is distributed. Show your calculations of the angles of the sectors of the pie chart. Use colouring or shading to emphasise how small Alvin's share is compared with what goes to the others.

(b) Alvin's LP and that of Donny Almond have been on sale in Disco Dan's shop for the past 8 weeks, during which time the number of records sold has been as follows:

	A. Yardbrush	D. Almond
Week 1	5	6
Week 2	12	58
Week 3	27	94
Week 4	53	100
Week 5	87	12
Week 6	82	5
Week 7	15	8
Week 8	12	2
Total	293	285

Plot two graphs on the same axis to illustrate these figures. Join your points by straight lines. Comment briefly on the main difference in the sales of the two records which are brought out by your graphs.

7. A survey of the type of holiday taken last year by the 30 pupils in form 4X gave the following information:

Holidays abroad	14 pupils
Seaside holiday in Great Britain	6 pupils
Other holiday in Great Britain	8 pupils
No holiday	2 pupils

Represent the above information as a pie chart using a circle of 6 cm radius. Carefully show any calculations that you do and mark on your diagram the sizes of the angles you have calculated.

8.

Population (millions)	
1961	1978
England 46.2	49.1
Scotland 5.2	5.2
N. Ireland 1.4	1.5

The populations of the United Kingdom in 1961 and 1978 are to be represented by two comparable pie charts.

(a) Calculate the angles at the centres of the sectors representing the populations of England, Scotland and Northern Ireland in 1961 giving your answers to the nearest degree.

(b) Using a circle of 5 cm radius, draw a pie chart for 1961.

(c) Calculate the radius of a comparable pie chart for 1978, giving your answer correct to three significant figures. Do not draw the pie chart for 1978.

9. A chemical company uses three raw materials, A, B and C in the manufacture of a certain product. The company's expenditure in thousands of pounds on these raw materials for the years 1976 – 79 are shown below. Construct a bar chart to illustrate the yearly expenditure and their constituent parts.

Expenditure (£thousands)			
	1976	1977	1978	1979
A	10	13	16	17
B	28	29	30	30
C	22	23	27	33

MULTIPLE CHOICE QUESTIONS

In the following write down the letter corresponding to the correct answer.

10. A pie chart was drawn to represent the following information:

Ingredient	Amount (grams)
Flour	280
Fruit	70
Egg	20
Butter	30
Sugar	100

The sector angle used to represent 'fruit' was

a 70°　　　　　　b 5°　　　　　　c 50°　　　　　　d 100°

11. A sphere of 3 cm radius is used to represent 27 000 tonnes. A sphere of 2 cm radius would represent

    a 24 000 t　　　　b 8000 t　　　　c 18 000 t　　　　d 12 000 t

12. A square of 4 cm side is used to represent 32 000 kg. A square of 2 cm side would represent

    a 16 000 kg　　　b 64 000 kg　　　c 8000 kg　　　　d 128 000 kg

13. In two comparable pie charts the radii are 5 cm and 8 cm. If the 5 cm circle represents a total expenditure of £50 000 then the 8 cm circle represents an expenditure of

    a £80 000　　　　b £31 250　　　　c £128 000　　　d £19 500

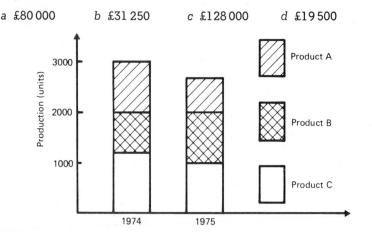

Fig. 1.

14. Fig. 1 shows that the production of product B was, in 1974,

    a 3000 units　　　b 1200 units　　　c 800 units　　　d 1000 units

15. The size of a farm is 120 hectares. It is cropped with fruit, cereals, potatoes and hay. The number of hectares used for each crop is shown on a pie chart and the sector angle representing cereals is 60°. What number of hectares are used for cereals?

    a 60　　　　　　b 30　　　　　　c 20　　　　　　d 90

　　　　　　　　　　　　　　　　　　　　　　A First Course in Statistics

3 DEGREES OF ACCURACY

3.1 COUNTING AND MEASURING

When data are obtained by counting, only one correct answer is possible. For instance, the number of marbles in a bag can be determined exactly.

However, in measuring anything, we can never find the exact measurement. The diameter of a ball-bearing might be found to be 2 mm by taking a rough measurement. If the diameter is measured more accurately it might be found to be 2.3 mm, correct to the nearest 0.1 mm. If we measure it with a more precise instrument (a micrometer) we might find the diameter to be 2.34 mm correct to the nearest 0.01 mm. The measurement of 2.34 mm is not the exact measurement any more than is 2.3 mm, although 2.34 mm is nearer to the correct diameter than is 2.3 mm.

Whenever a measurement is taken — whether it be a mass, a length, a period of time or a scientific measurement — we are limited by the accuracy of the equipment used and by our own human limitations. The exact measurement can never be found and hence we have to be content with an approximation.

3.2 APPROXIMATE NUMBERS

One way of obtaining an approximate number is to state it to so many significant figures. The rules are as follows:

(i)   If the first figure to be discarded is less than 5 do not alter the previous figure. Thus

$$74\,342 = 74\,340 \text{ correct to 4 significant figures}$$
$$= 74\,300 \text{ correct to 3 significant figures}$$
$$= 74\,000 \text{ correct to 2 significant figures.}$$

(ii)   If the first figure to be discarded is more than 5 increase the previous figure by one. Thus

$$87\,867 = 87\,870 \text{ correct to 4 significant figures}$$
$$= 87\,900 \text{ correct to 3 significant figures}$$
$$= 88\,000 \text{ correct to 2 significant figures.}$$

(iii) When the first figure to be discarded is 5, if the number to the left of the 5 is odd then do not alter the previous number. If the number to the left of the 5 is even then increase the previous number by one.

3475 = 3470 correct to 3 significant figures
3485 = 3490 correct to 3 significant figures.

If two or more digits are to be discarded, and the first of these is 5, then the last remaining digit should be increased by 1.

24 653 = 24 700 correct to 3 significant figures
0.075 549 = 0.076 correct to 2 significant figures.

(iv) Zeros must be kept to show the position of the decimal point or to indicate that the zero is a significant figure.

14 384   = 14 400 correct to 3 significant figures
0.0883   = 0.088 correct to 2 significant figures (note that the zero following the decimal point is not regarded as being a significant figure)
0.007 365 = 0.007 correct to 1 significant figure
29. 8604   = 29.860 correct to 5 significant figures.

A second way of obtaining an approximate number is to state the number to so many decimal places. The number of figures after the decimal point determines the number of decimal places. Thus

0.0493 = 0.049 correct to 3 decimal places
      = 0.05 correct to 2 decimal places.
11.273 = 11.27 correct to 2 decimal places
      = 11.3 correct to 1 decimal place.

EXERCISE 4

1. Write down the following numbers correct to the number of significant figures stated:
   (a) 24.865 82   (i) to 4   (ii) to 2   (iii) to 6
   (b) 0.008 357 1   (i) to 4   (ii) to 2   (iii) to 3
   (c) 4.978 48   (i) to 5   (ii) to 3   (iii) to 1
   (d) 21.987   to 2
   (e) 35.603   to 4
   (f) 28 387 617   (i) to 5   (ii) to 2
   (g) 4.149 75   (i) to 5   (ii) to 4   (iii) to 3
   (h) 9.2045   (i) to 4   (ii) to 3

2. Write down the following numbers correct to the number of decimal places stated:
   (a) 5.149 87   (i) to 4   (ii) to 3   (iii) to 2
   (b) 35.2856   (i) to 2   (ii) to 3   (iii) to 1
   (c) 0.004 977   (i) to 5   (ii) to 4   (iii) to 3
   (d) 8.4076   (i) to 3   (ii) to 2   (iii) to 1
   (e) 0.8529   (i) to 3   (ii) to 2

3.3 ROUNDING

When a number is stated to so many significant figures we say that it has been *rounded* to so many significant figures. Thus the number

186 749 = 187 000 rounded to 3 significant figures.

3.4 BIASED ROUNDING

When the rounding is performed according to the rules given above the errors occurring are called *unbiased* or *compensating errors.*

If the rounding takes place in one direction only, as shown in the table below, bias arises. The answers obtained by adding the numbers given in column 1 are much less accurate in columns 2 and 3 than they are in column 4, where the rounding has been performed correctly.

Column 1	Column 2	Column 3	Column 4
Given numbers	Rounded down to lower hundred	Rounded up to upper hundred	Rounded correctly to 3 sig. figures
19 632	19 600	19 700	19 600
27 581	27 500	27 600	27 600
30 037	30 000	30 100	30 000
24 569	24 500	24 600	24 600
18 775	18 700	18 800	18 800
120 594	120 300	120 800	120 600

3.5 MAXIMUM ERROR IN A ROUNDED NUMBER

Suppose that the number 380 has been rounded correct to 2 significant figures. The original number, from which 380 was obtained, could have been as large as 385 or as small as 375. That is, the number from which 380 was obtained must lie within the limits 380 ± 5. The number 5 is said to be the *absolute error.*

In the case of measurements the absolute error is half of the smallest unit of measurement.

EXAMPLE 1. A period of time was measured as 28.7 seconds. State

    (a) the smallest unit of measurement

(b) the aboslute error

(c) the limits within which the true measurement must lie.

    (a) The smallest unit of measurement is 0.1 second.

(b) The absolute error is half of 0.1 second = 0.05 second.

(c) The limits are 28.7 ± 0.05 second. That is, the upper limit is 28.7 + 0.05 = 28.75 seconds and the lower limit is 28.7 − 0.05 = 28.65 seconds.

EXAMPLE 2. A distance is measured as 500 metres correct to the nearest 10 metres. The time taken by a vehicle to travel this distance is measured as 9.8 seconds correct to the nearest 0.1 second. Find the greatest and least possible speed of the vehicle.

$$\text{Speed} = \frac{\text{distance}}{\text{time}}$$

$$\text{Greatest possible speed} = \frac{\text{longest possible distance}}{\text{least possible time}}$$

$$= \frac{505}{9.75} = 51.8 \, \text{m/s}$$

$$\text{Least possible speed} = \frac{\text{shortest possible distance}}{\text{greatest possible time}}$$

$$= \frac{495}{9.85} = 50.3 \, \text{m/s}$$

3.6 RELATIVE ERROR

The relative error is often a more useful measure of accuracy than is the absolute error. It is found by the following formula:

$$\text{Relative error} = \frac{\text{absolute error}}{\text{true value}}$$

If the true value is not known, as will often be the case, the true value may be taken as being the approximate value. Relative errors are frequently expressed as percentages. Thus

$$\% \text{ relative error} = \frac{\text{absolute error}}{\text{true value}} \times 100$$

                                                 A First Course in Statistics

EXAMPLE 3. Find the relative error in measuring a length as 8.6 cm, correct to the nearest 0.1 cm.

Smallest unit of measurement $= 0.1$ cm

Absolute error $= 0.05$ cm

% relative error $= \dfrac{0.05}{8.6} \times 100 = 0.58\%$.

3.7   ADDING AND SUBTRACTING ROUNDED NUMBERS

EXAMPLE 4.   Determine the upper and lower limits of the sum of 450, 560, 770 and 310, each number being given correct to 2 significant figures.

Rounded number	Absolute error	Greatest possible value	Least possible value
450	± 5	455	445
560	± 5	565	555
770	± 5	775	765
310	± 5	315	305
2090	± 20	2110	2070

Hence the upper limit is $2090 + 20 = 2110$ and the lower limit is $2090 - 20 = 2070$.

The sum of a set of rounded numbers is equal to the sum of the rounded numbers plus or minus the sum of the absolute error in each number.

EXAMPLE 5. Find the greatest and least possible values of the sum of 8340 and 9680, each number being rounded correct to 3 significant figures.

$8340 + 9680 = 18\,020$

Absolute error in each number $= \pm 5$

Sum of the absolute errors $= \pm 5 \pm 5 = \pm 10$

Greatest possible value $= 18\,020 + 10 = 18\,030$

Least possible value $= 18\,020 - 10 = 18\,010$.

EXAMPLE 6. The numbers 938 and 762 have been rounded correct to 3 significant figures. Find the greatest and least values of their difference.

Rounded number	Limits of original numbers	
938 ± 0.5	938.5 (highest)	937.5 (lowest)
762 ± 0.5	761.5 (lowest)	762.5 (highest)
176 ± 1	177	175

Greatest possible value $= 176 + 1 = 177$

Least possible value $= 176 - 1 = 175$

Note that to find the absolute error in the difference we still add the individual errors.

If rounded numbers are to be added or subtracted they should be rounded to the same number of significant figures.

3.8   PRODUCTS AND QUOTIENTS

EXAMPLE 7. **Multiply 2.16 and 3.28, each number having been rounded correct to** 3 significant figures.

Product of the rounded numbers $= 2.16 \times 3.28 = 7.0848$

Greatest possible product $= 2.165 \times 3.285 = 7.112\,025$

Least possible product $= 2.155 \times 3.275 = 7.057\,625$

We see that it is not possible to state the product of the two numbers more accurately than 7.1 correct to 2 significant figures.

Generally an answer should not contain more significant figures than the least number of significant figures amongst the given numbers.

EXAMPLE 8. Find the value of $\dfrac{0.7283 \times 0.073\,58}{0.0384}$ each number having been rounded correct to the number of significant figures shown.

Using a calculator would show that

$$\frac{0.7283 \times 0.073\,58}{0.0384} = 1.395\,529\,01$$

However, the least number of significant figures amongst the given numbers is three (for the number 0.0384). Hence the answer should only be stated to three significant figures, that is as 1.40.

EXERCISE 5

Find the lower and upper limits for each of the following calculations:

1. 432 + 768

2. 5630 + 2640 + 3980 + 7680

3. 8.65 + 7.32 + 1.68

4. 75 − 68

5. 93 000 − 68 000

6. 1156 − 1068

7. 18 × 23

8. 2.06 × 9.17

9. 11 000 × 26 000

10. 6.08 ÷ 1.29

11. 16 ÷ 70

12. 15 600 ÷ 11 300

13. Find the absolute error in measuring a length of 11.5 cm correct to the nearest 0.1 cm.

14. Find the absolute error and the relative percentage error in measuring a period of time as 362 seconds correct to the nearest second.

15. Find the maximum relative error in the product of 18 × 32, both numbers being correct to 2 significant figures.

16. If each of the numbers in the fraction below is correct to two significant figures, calculate the greatest and least values of the fraction.

$$\frac{9.6 - 5.3}{7.1 - 6.0}$$

17. Correct to the nearest whole number, the value of x is 2 and the value of y is 1. Calculate the greatest possible value for each of the following algebraic expressions.

(a) $\dfrac{x}{y}$

(b) $\dfrac{y}{7y - x}$

18. In each of the following every number is correct to the number of significant figures shown. Find the value of each of them stating the answer to an appropriate number of significant figures.

(a) 223.6×0.0048

(b) 32.7×0.259

(c) $0.682 \times 0.097 \times 2.38$

(d) $78.41 \div 23.78$

(e) $0.059 \div 0.00268$

(f) $33.2 \times 29.6 \times 0.031$

(g) $\dfrac{0.728 \times 0.006\,25}{0.0281}$

(h) $\dfrac{27.5 \times 30.52}{11.3 \times 2.73}$

19. A man runs 200 m in 24 seconds. If the distance is accurate to the nearest 10 m and the time is accurate to 0.1 second, find the greatest possible speed and the slowest possible speed.

4 FREQUENCY DISTRIBUTIONS

4.1 RAW DATA

Raw data are collected data which have not yet been arranged in any sort of order.

Consider the marks of 50 students obtained in a test (marks out of 10):

4	3	5	4	3	5	5	4	3
6	5	4	5	4	5	3	4	4
5	5	7	4	3	4	3	4	5
4	3	6	1	3	6	3	2	6
6	3	5	2	7	5	7	1	7
6	5	8	6	4				

This is an example of raw data where we can see that the numbers have not been arranged in any systematic way.

One way of organising raw data into order is to form a frequency distribution. The number of students scoring 3 marks is found, the number scoring 4 marks and so on. A tally chart is the best way of doing this.

On looking at the raw data we see that the smallest mark is 1 and that the greatest mark is 8. The numbers from 1 to 8 are written in column 1 of the tally chart (Table 4.1). We now take each figure in the raw data in turn and for each figure we place a tally mark opposite the appropriate figure.

The fifth tally mark for each number is usually made in an oblique direction thereby tying the tally marks into bundles of five.

When the tally marks are complete they are counted and the numerical value recorded in the column headed 'frequency'. Hence the frequency is the number of times each mark occurs. From the tally chart it will be seen that the mark 1 occurs twice (a frequency of 2), the mark 5 occurs twelve times (a frequency of 12) and so on.

Table 4.1.

Mark	Tally	Frequency
1	||	2
2	||	2
3	ЖН ЖН	10
4	ЖН ЖН ||	12
5	ЖН ЖН ||	12
6	ЖН ||	7
7	||||	4
8	|	1
	Total	50

4.2   GROUPED FREQUENCY DISTRIBUTIONS

When dealing with a large amount of data it is useful to group the information into classes or categories. We can then determine the number of items which belong to each class thereby obtaining a class frequency as shown in Example 1.

EXAMPLE 1. The numbers shown below are the times in seconds (to the nearest second) for 40 children to complete a length of a swimming pool. The swimmers were divided into heats as the pool had eight lanes.

	Lane number							
Heat	1	2	3	4	5	6	7	8
I	40	49	43	35	· 42	43	46	36
II	42	36	37	44	39	41	31	45
III	38	48	44	51	38	53	35	32
IV	30	43	41	52	46	43	50	40
V	39	41	48	47	32	52	47	42

Display these data in the form of a grouped frequency table using intervals 30 − 34, 35 − 39, etc.

In order to obtain the grouped frequency distribution we use a tally chart as shown in Table 4.2

Table 4.2.

Class	Tally	Frequency
30 – 34	IIII	4
35 – 39	JHf IIII	9
40 – 44	JHf JHf IIII	14
45 – 49	JHf III	8
50 – 54	JHf	5
	Total	40

The main advantage of grouping is that it produces a clear overall picture of the distribution. However, too many groups will destroy the pattern of the distribution whilst too few will destroy much of the detail which was present in the raw data. Depending upon the volume of raw data the number of classes is usually between 5 and 20.

4.3   CLASS INTERVALS

In Table 4.2, the first class is 30 – 34. These figures give the class interval. For the second class, the class interval is 35 – 39. The end numbers 35 and 39 are called the *class limits* for the second class, 35 being the lower class limit and 39 the upper class limit.

4.4   CLASS BOUNDARIES

In Table 4.2, the times have been recorded to the nearest second. The class interval 35 – 39 theoretically includes all the times between 34.5 seconds and 39.5 seconds. These numbers are called the *lower and upper class boundaries* respectively for the second class. For any distribution, the class boundaries may be found by adding the upper class limit of one class to the lower class limit of the next class and dividing this sum by two.

EXAMPLE 2. The figures below show part of a frequency distribution for the life-times of electric light bulbs. State the lower and upper class boundaries for the second class.

Lifetime (hours)	Frequency
400 – 449	22
450 – 499	38
500 – 549	62

For the second class:

$$\text{Lower class boundary} = \frac{449 + 450}{2} = 449.5 \text{ hours}$$

$$\text{Upper class boundary} = \frac{499 + 500}{2} = 499.5 \text{ hours}$$

4.5   WIDTH OF A CLASS INTERVAL

The width of a class interval is the difference between the lower and upper class boundaries. That is:

Width of class interval = upper class boundary − lower class boundary

For Example 2:

Width of second class interval = 499.5 − 449.5 = 50 hours

(A common mistake is to take the class width as being the difference between the upper and lower class limits, giving in Example 1, 499 − 450 = 49 hours, which is incorrect.)

4.6   DISCRETE AND CONTINUOUS VARIABLES

A variable which can, theoretically, take any value between two given values is called a *continuous variable*. Thus the height of an individual which can be 158 cm, 164.2 cm, or 177.832 cm, depending upon the accuracy of measurement, is a continuous variable.

A variable which can only take certain values is a *discrete variable*. Thus the number of children in a family can only take the values 0, 1, 2, 3, ..., but it cannot be 2½, 3¼, etc. It is therefore a discrete variable. Note that the values of a discrete variable need not be whole numbers. The size of shoes is a discrete variable but these can be 4½, 5, 5½, 6, etc.

4.7 HISTOGRAMS

The histogram is a diagram which is used to represent a frequency distribution. It consists of a set of rectangles whose *areas* represent the frequencies of the various classes. If all the classes have the same width then all the rectangles will have the same width and the frequencies are then represented by the heights of the rectangles. Fig. 4.1 shows the histogram for the frequency distribution of Table 4.1.

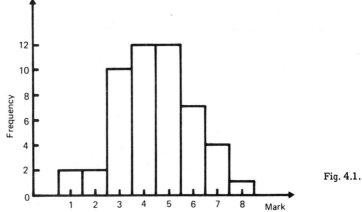

Fig. 4.1.

4.8 HISTOGRAM FOR A GROUPED DISTRIBUTION

A histogram for a grouped distribution may be drawn by using the mid-points of the class intervals as the centres of the rectangles. The histogram for the distribution of Table 4.2 is shown in Fig. 4.2. Note that the extremes of the base of each rectangle represent the lower and upper class boundaries.

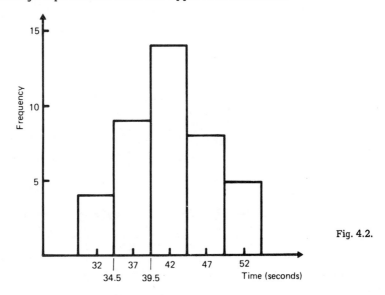

Fig. 4.2.

Frequency Distributions

EXAMPLE 3. The table below shows the age distribution for workers in a certain firm.

Age group	16–20	21–25	26–30	31–40	41–50	51–70
Number of workers	20	18	14	18	16	24

Draw a histogram of this information.

The first thing to notice is that the classes are not all the same width and in drawing the histogram we need to remember that it is the *areas of the rectangles* which give the frequencies of the various classes. Let us examine the widths of the classes:

Age group	16–20	21–25	26–30	31–40	41–50	51–70
Class width	5	5	5	10	10	20

In drawing the histogram (Fig. 4.3) let us take 1 unit to represent a class width of 5 years. Then a class width of 10 years will be represented by a rectangle with a width of 2 units and a class width of 20 years will be represented by a rectangle with a width of 4 units.

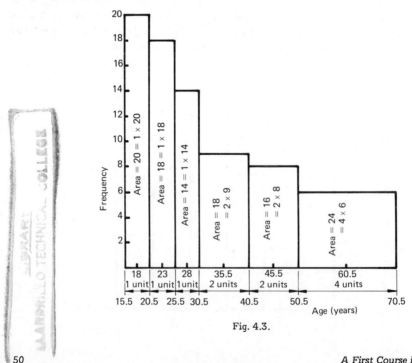

Fig. 4.3.

The heights of the rectangles then become:

Age group	16 – 20	21 – 25	26 – 30	31 – 40	41 – 50	51 – 70
Height of rectangle	$\frac{20}{1} = 20$	$\frac{18}{1} = 18$	$\frac{14}{1} = 14$	$\frac{18}{2} = 9$	$\frac{16}{2} = 8$	$\frac{24}{4} = 6$

EXAMPLE 4. Fig. 4.4 shows a histogram for the lifetime of electric bulbs. Draw up a frequency distribution.

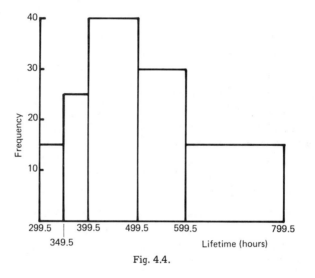

Fig. 4.4.

Classes 1 and 2 are 1 unit wide (1 unit being 50 hours) classes 3 and 4 are 2 units wide whilst class 5 is 4 units wide. From the histogram the frequencies are:

Class 1	15	Class 4	$30 \times 2 = 60$
Class 2	25	Class 5	$15 \times 4 = 60$
Class 3	$40 \times 2 = 80$		

The frequency distribution is:

Lifetime (hours)	300 – 349	350 – 399	400 – 499	500 – 599	600 – 799
Frequency	15	25	80	60	60

4.9   OPEN-ENDED CLASSES

The information shown in the table below gives the distribution of weekly incomes within an industrial organisation.

Income (£)	Number of people
Less than 50	5
51 – 60	12
61 – 70	18
71 – 80	25
81 – 90	17
91 – 100	8
Over 100	3

The first and last classes are called open-ended classes because in the case of the first class no lower limit is stated and in the case of the last class no upper limit is given. Classes are left open-ended for many reasons, but usually, because to specify a limit might be misleading. Thus in the case of the last class 'over 100' the highest income might be paid to the managing director and it might be in the region of £400 per week whilst the next highest income might be paid to a works manager at £200 per week.

When drawing the histogram of an open-ended distribution, some attempt must be made to state the missing limits. In our example the first class will be taken as 41 – 50 and the last class as 101 – 110. The histogram can then be drawn in the usual way.

4.10 DISCRETE DISTRIBUTIONS

The histograms drawn so far have all represented distributions in which the variable was continuous. The data in the next example are discrete and we shall see how this kind of information is represented.

EXAMPLE 5. 5 coins were tossed 100 times and after each toss the number of heads was recorded. The table on the following page gives the number of tosses during which 0, 1, 2, 3, 4 and 5 heads were obtained. Represent the data in a suitable diagram.

Number of heads	Number of tosses (frequency)
0	4
1	15
2	34
3	29
4	16
5	2
	Total 100

Since the data are discrete (there cannot be 2.3 or 3.6 heads) Fig. 4.5(a) seems the most natural diagram to use. This diagram is in the form of a vertical bar chart in which the bars have zero width. Fig 4.5(b) shows the same data represented as a histogram. Note that the area under the diagram gives the total frequency of 100 which is as it should be. Discrete data are often represented as a histogram, although in doing this we are treating the data as though they were continuous.

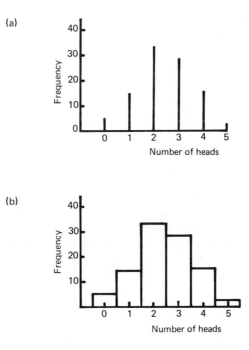

Fig. 4.5. Histograms representing discrete data

EXERCISE 6

1. The following marks were obtained by 50 students during a test:

4	3	5	4	3	5	5	4	3	6	5	4	5	3	4	4	5	5	7	4	3
4	3	4	5	4	3	6	1	3	6	3	2	6	6	3	5	2	7	5	7	1
7	6	5	8	6	4	3	5													

    Obtain a frequency distribution by means of a tally chart, and hence draw a histogram of this information.

2. A survey was made one evening of the ages of 30 members of a youth club, with the following results:

14	16	16	15	15	14
14	17	18	15	14	16
16	16	14	14	15	17
15	14	13	14	14	18
16	15	14	13	17	17

    Tabulate the above results in a suitable frequency distribution table and draw a histogram to illustrate your results.

3. In a swimming match, the times taken, to the nearest second, by 20 children to swim a length were as follows:

31	27	24	26	31	25	26	32	27	31
26	32	30	32	29	25	29	27	26	32

    Obtain a frequency distribution and hence draw a histogram of this information.

4. In the game of bridge, one method of evaluating the strength of a hand is to count 4 points for each Ace, 3 points for a King, 2 points for a Queen, and 1 point for a Jack. A player keeps a record of the points he received over 50 deals. They are as follows:

16	1	17	8	9	11	19	13	10	18
21	2	13	10	14	15	10	4	11	20
12	10	28	9	11	12	9	11	9	5
6	16	8	24	7	9	8	7	6	0
3	7	12	10	10	13	5	15	14	8

(a) Using these figures copy out and complete in full the following table:

No. of points	Tally	Frequency
0 – 3		
4 – 7		
8 – 11		
12 – 15		
16 – 19		
20 – 23		
24 – 27		
28 – 31		

(b) Draw a histogram of this information.

5.   The following is a record of the percentage marks obtained by 100 students in an examination.

```
45   93  35  56  16  50  63  30  86  65  57  39  44  75  25  45  74
93   84  25  77  28  54  50  12  85  55  34  50  57  55  48  78  15
27   79  68  26  66  80  91  62  67  52  50  75  96  36  83  20  45
71   63  51  40  46  61  62  67  57  53  45  51  40  46  31  54  67
66   52  49  54  55  52  56  59  38  52  43  55  51  47  54  56  56
42   53  40  51  58  52  27  56  42  86  50  31  61  33  36
```

Draw up a tally chart for the classes 0 – 9, 10 – 19, 20 – 29, ..., 90 – 99, and hence form a frequency distribution and use this to draw a histogram.

6.   Draw a histogram of the following data which relates to the ages of children in a small school.

Age (years)	11	12	13	14	15	16
Frequency	42	53	57	52	50	41

7. The histogram (Fig. 4.6) illustrates the frequency distribution of the heights (in centimetres) of a group of 54, 14 year old boys.

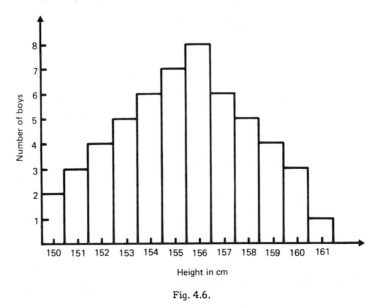

Fig. 4.6.

Copy and complete the following table:

Height (cm)	150 151 152 153 154 155 156 157 158 159 160 161
Frequency	

8. Draw a histogram of the following data:

Length (nearest cm)	11 – 15	16 – 20	21 – 25	26 – 30	31 – 35
Frequency	4	7	12	8	3

9. State which of the following represent discrete data and which represent continuous data.

   (a) Rainfall in a city during various months of the year.

   (b) Speed of a car in m s⁻¹.

   (c) Number of five pound notes circulating in England at any time.

   (d) Number of goals scored in a game of soccer.

10. Part of a frequency distribution table is shown below.

Length (cm)	200 – 209	210 – 229	230 – 249
Frequency	40		

On the histogram representing this distribution the height of the column representing 200 – 209 was 5 cm. If the height of the column corresponding to 210 – 229 was 4 cm, calculate the frequency.

11. The data below give the diameters of machined parts

Diameter (mm)	14.96 – 14.98	14.99 – 15.01	15.02 – 15.04 etc.
Frequency	3	8	12

Write down:

(a) the upper and lower class boundaries for the second class

(b) the class width of the classes shown in the table.

12. Classify each of the following as continuous or discrete variables:

(a) the diameters of ball bearings

(b) the number of shirts sold per day

(c) the mass of packets of chemicals

(d) the number of bunches of daffodils packed by a grower

(e) the lifetime of electric bulbs.

13. A batch of 100 samples of crockery showed the following frequency distribution of flaws.

No. of flaws	Less than 3	3 – 5	6 – 8	9 – 11	12 – 14	15 – 17	18 and over
Frequency	10	22	30	21	12	4	1

Draw a histogram of this information.

14. Draw a histogram to represent the data below:

Height (cm)	0 – 9	10 – 19	20 – 29	30 – 49	50 –
Frequency	40	32	48	56	60

15. The weekly earnings of men employed in a certain company were as follows:

Weekly earnings	Number
Under £40	3
£40 but under £44	4
£44 but under £50	14
£50 but under £60	43
£60 but under £70	74
£70 but under £80	61
£80 but under £90	178
£90 but under £100	86
£100 but under £120	164
£120 but under £160	103
£160 and over	93

Draw a histogram to represent these data.

16. The information below is taken from a histogram.

Length (m)	10 – 19	20 – 39	40 – 59	60 – 99
Height of rectangle (cm)	3	5	6	2

The frequency of the first class is 15. Determine the frequencies of the remaining classes.

4.11 FREQUENCY POLYGONS

The frequency polygon provides a second way of representing a frequency distribution. It is drawn by connecting the mid-points of the tops of the rectangles in the histogram by straight lines.

EXAMPLE 6. Draw a frequency polygon for the information given below:

Age of employees	15 – 19	20 – 24	25 – 29	30 – 34	35 – 39
Number	5	23	58	104	141

Age of employees	40 – 44	45 – 49	50 – 54	55 – 59
Number	98	43	19	6

Draw a frequency polygon to represent these data.

The frequency polygon is drawn in Fig. 4.7. It is customary to add the extensions PQ and RS to the next lower and next higher class mid-points as shown in the diagram. If this is done, the area of the polygon is equal to the area of the histogram.

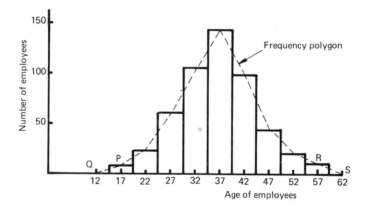

Fig. 4.7.

4.12 FREQUENCY CURVES

Collected data can be considered as forming a sample which has been taken from a large population. Very many observations are possible in the population and hence, for continuous data, we can choose very small class intervals and still have reasonable numbers of observations falling within each class. The rectangles making up the histogram then become very small in width and the frequency polygon, to all intents, becomes a curve. This curve is called a *frequency curve*.

4.13 TYPES OF FREQUENCY CURVE

The frequency curves which occur in practice take on certain characteristic shapes as shown in Fig. 4.8.

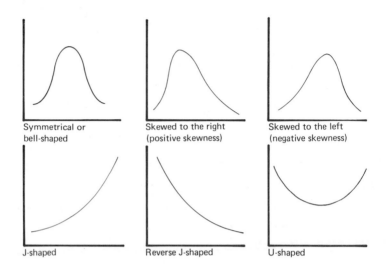

Fig. 4.8. Types of frequency curves.

4.13.1 *Symmetrical or Bell-shaped Distributions (usually called normal curves).*

These occur when measurements e.g. the heights of individuals in a human population or the intelligence quotients of individuals.

4.13.2 *Moderately Skewed Distributions*

These often occur when a small sample is taken from a population which would give a symmetrical distribution. For instance, if a sample of 50 men are measured for height the distribution will, almost certainly, be skewed. However, many positively skewed distributions occur in their own right. Some examples are: the number of children per family, the age at which women marry and the distribution of income in the UK. Negatively skewed distributions occur very rarely and an investigator faced with this type of distribution would suspect bias of some sort.

4.13.3 *J—type Distribution*

The positive type (reverse J—shaped) is much more common than the negative type. J—type distributions occur, for instance, for the lifetimes of electronic devices, and for the parking times of vehicles in a street.

A First Course in Statistics

4.13.4 U—shaped Distributions

Very few examples of U—shaped distributions occur in statistics. An example is provided by the number of cars on the roads during the day. During the morning rush-hour the number of cars is very large but the number of cars falls off until a minimum is reached at some time before lunch-time. The number of cars then starts to rise again until the evening rush-hour occurs. However, this type of curve also occurs when a minimum value is being sought. For instance in trying to determine the minimum cost of producing an article.

4.13.5 Bimodal Distributions

These are distributions with two peaks. Frequently a bimodal distribution occurs because of a mixture of two symmetrical distributions (Fig. 4.9).

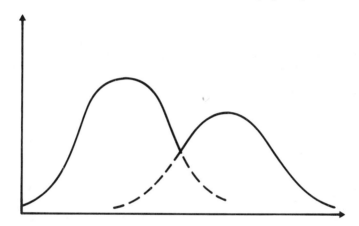

Fig. 4.9. A bimodal distribution arising from a mixture of two symmetrical distributions.

4.14 CUMULATIVE FREQUENCY DISTRIBUTIONS

A cumulative frequency distribution is an alternative method of presenting a frequency distribution. The way in which a cumulative frequency distribution is obtained is shown in Example 7.

EXAMPLE 7. Obtain a cumulative frequency distribution for the data given in the table below.

Height (cm)	Frequency
150 – 154	8
155 – 159	16
160 – 164	43
165 – 169	29
170 – 174	4

The class boundaries are 149.5 − 154.5, 154.5 − 159.5, 159.5 − 164.5, 164.5 − 169.5 and 169.5 − 174.5. In drawing up a cumulative frequency distribution the lower boundary limit for each class is used.

Height (cm)	Cumulative frequency
Less than 149.5	0
Less than 154.5	8
Less than 159.5	8 + 16 = 24
Less than 164.5	24 + 43 = 67
Less than 169.5	67 + 29 = 96
Less than 174.5	96 + 4 = 100

The distribution may be represented by a cumulative frequency polygon (Fig. 4.10) or by a cumulative frequency curve (Fig. 4.11) known as an *ogive* (after the architechtural term used for this type of shape).

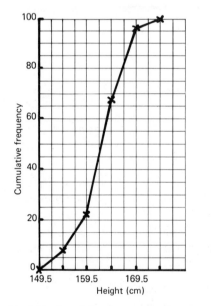

Fig. 4.10. Cumulative frequency polygon.

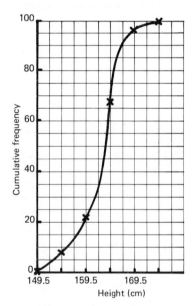

Fig. 4.11. Cumulative frequency curve or ogive.

The cumulative distribution shown above is called a 'less than' cumulative frequency distribution and when a cumulative distribution is required this is the one commonly used. However, for some purposes, a 'more than' distribution is needed and this is obtained as shown opposite.

Height (cm)	Cumulative frequency
More than 149.5	100
More than 154.5	$100 - 8 = 92$
More than 159.5	$92 - 16 = 76$
More than 164.5	$76 - 43 = 33$
More than 169.5	$33 - 29 = 4$
More than 174.5	0

The cumulative frequency curve for this 'more than' distribution is shown in Fig. 4.12.

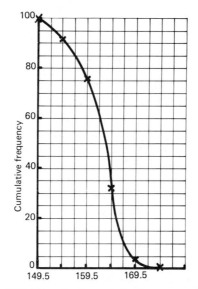

Fig. 4.12. 'More than' cumulative frequency curve.

EXAMPLE 8. The diameters of 200 ball-bearings were measured with the results as shown below:

Diameter (mm)	5.94 – 5.96	5.97 – 5.99	6.00 – 6.02	6.03 – 6.05	6.06 – 6.08
Number	8	37	90	52	13

Draw a cumulative frequency curve for this information and from it estimate the number of ball-bearings

(a) with a diameter less than 5.98 mm

(b) with a diameter between 6.00 mm and 6.05 mm

(c) with a diameter greater than 6.07 mm.

The cumulative frequencies are as follows:

Diameter (mm)	Cumulative frequency
Less than 5.965	8
Less than 5.995	45
Less than 6.025	135
Less than 6.055	187
Less than 6.085	200

The ogive is shown in Fig. 4.13 where it will be seen that:

(a) The cumulative frequency corresponding to a diameter of 5.98 mm is 22. Hence 22 ball-bearings have a diameter less than 5.98 mm.

(b) The cumulative frequencies corresponding to 6.00 mm and 6.05 mm are 64 and 177 respectively. Hence the number of ball-bearings having a diameter between 6.00 mm and 6.05 mm is $177 - 64 = 113$.

(c) The cumulative frequency corresponding to a diameter of 6.07 mm is 195. Hence the number of ball-bearings with a diameter greater than 6.07 mm is $200 - 195 = 5$.

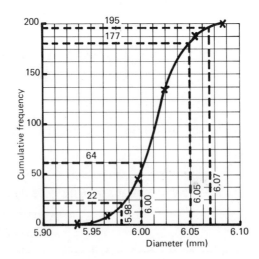

Fig. 4.13.

EXAMPLE 9. Sixty children in a certain school have a test marked out of 70 marks. The cummulative frequency curve is shown in Fig. 4.14. Copy and complete the table below:

Mark	0 – 9	10 – 19	20 – 29	30 – 39	40 – 49	50 – 59	60 – 69
Frequency							

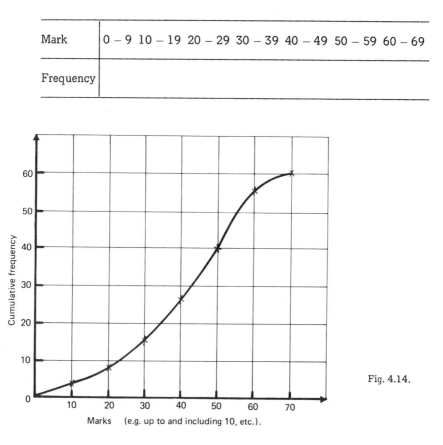

Fig. 4.14.

Using the curve gives the following cumulative frequencies:

Mark	Cumulative frequency
Less than 10	4
Less than 20	8
Less than 30	15
Less than 40	27
Less than 50	40
Less than 60	55
Less than 70	60

The frequency distribution is then:

Mark	Frequency
0 – 9	4
10 – 19	8 – 4 = 4
20 – 29	15 – 8 = 7
30 – 39	27 – 15 = 12
40 – 49	40 – 27 = 13
50 – 59	55 – 40 = 15
60 – 69	60 – 55 = 5

Hence the completed table is:

Mark	0 – 9	10 – 19	20 – 29	30 – 39	40 – 49	50 – 59	60 – 69
Frequency	4	4	7	12	13	15	5

4.15 PERCENTAGE OGIVES

If percentage cumulative frequencies are used to draw the ogive the diagram is called a percentage ogive. Percentage ogives are useful when two or more frequency distributions are to be compared.

EXAMPLE 10. The information given in the table below gives the weekly wages paid by two firms A and B.

Weekly wage (£)	Firm A	Firm B
50 – 59.99	16	24
60 – 69.99	20	25
70 – 79.99	34	48
80 – 89.99	30	37
90 – 99.99	20	23
100 – 109.99	10	12
110 – 119.99	3	5
Totals	133	174

Draw a percentage ogive for each distribution using the same axes and scales.

Strictly we should use the lower boundary limits for each class when drawing the percentage ogive but no great error will result if we use £60, £70, £80, etc. instead.

Weekly Wage (£)	Firm A		Firm B
	Cumulative frequency	% Cumulative frequency	% Cumulative frequency
Less than 60	16	$\dfrac{16}{133}$ X 100 = 12.0	$\dfrac{24}{174}$ X 100 = 13.8
Less than 70	36	$\dfrac{36}{133}$ X 100 = 27.1	$\dfrac{49}{174}$ X 100 = 28.2
Less than 80	70	$\dfrac{70}{133}$ X 100 = 52.6	$\dfrac{97}{174}$ X 100 = 55.7
Less than 90	100	$\dfrac{100}{133}$ X 100 = 75.2	$\dfrac{134}{174}$ X 100 = 77.0
Less than 100	120	$\dfrac{120}{133}$ X 100 = 90.2	$\dfrac{157}{174}$ X 100 = 90.2
Less than 110	130	$\dfrac{130}{133}$ X 100 = 97.7	$\dfrac{169}{174}$ X 100 = 97.1
Less than 120	133	100.0	100.0

The ogives are shown in Fig. 4.15 and a comparison between the weekly wages paid by the two firms is now easy to make.

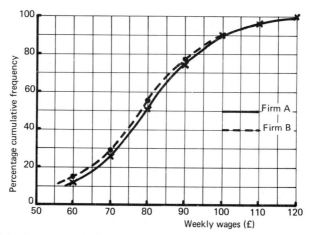

Fig. 4.15. Percentage ogives used for comparing two frequency distributions.

EXERCISE 7

1. In a survey of the value of orders received a manufacturing company obtained the following grouped frequency table:

Value of orders (£)	Number of orders
100 and up to 200	169
200 and up to 300	176
300 and up to 400	75
400 and up to 500	32
500 and up to 600	8

Draw the histogram for this distribution and also a frequency polygon.

2. (a) What form of distribution is shown by a frequency curve of Fig. 4.16?

    (b) Name *one* property of this distribution.

    (c) Name *two* populations which have this distribution.

Fig. 4.16.

3. The table below shows the age distribution of those males receiving retirement pensions at the end of a certain year.

Age (years)	Number receiving pensions (thousands)
65 – 69	1030
70 – 74	820
75 – 79	460
80 – 84	240
85 – 89	80
90 and over	20

Draw a frequency curve for these data and state its shape.

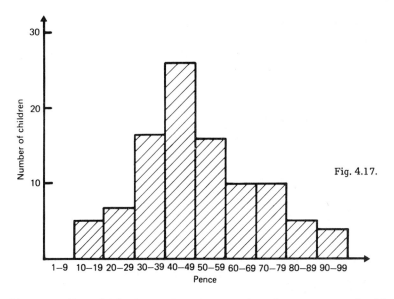

Fig. 4.17.

4. The histogram (Fig. 4.17) shows the amounts of pocket money received by a group of 12 year old children.

(a) Use the histogram to complete the table for the distribution:

Pocket money (p)	1 – 9	10 – 19	20 – 29	30 – 39	40 – 49
Number of children	0	5	7		

Pocket money (p)	50 – 59	60 – 69	70 – 79	80 – 89	90 – 99
Number of children					

(b) Complete the cumulative frequency table and use it to draw a cumulative frequency diagram.

Pocket money (p) Less than	10 20 30 40 50 60 70 80 90 100
Number of children	0 5 12 29

(c) From your cumulative frequency diagram estimate as closely as possible the number of children who receive between 34 p and 64 p pocket money.

5. Make up simple examples to illustrate each of the following:
   (a) a pictogram,

   (b) a time series graph,

   (c) a frequency polygon.

   Give the name of a diagram that could be used as an alternative in part (c).

6. The figures below are measurements of the noise level in (dBA units) at 36 'discos'.

93	90	98	88	103	92	89	82	86
87	91	89	85	95	86	86	94	85
103	92	88	102	99	85	98	100	95
98	105	86	100	92	96	91	87	88

   Display the data in the form of a grouped frequency table, using intervals 81 – 85, 86 – 90, etc. Draw a cumulative frequency polygon for your grouped frequency distribution.

   From your graph, estimate the percentage of 'discos' at which the noise level exceeds 94 dBA units.

7. A school decided that all 900 pupils should sit an intelligence test. The following table of results was obtained.

Range of mark	Frequency
0 – 9	5
10 – 19	33
20 – 29	82
30 – 39	148
40 – 49	270
50 – 59	215
60 – 69	87
70 – 79	44
80 – 89	12
90 – 99	4

   (a) (i) Construct a cumulative frequency table.
       (ii) Plot a cumulative frequency curve on graph paper. Use the longer side of the graph paper for the horizontal axis. Use 2 cm to 10 marks horizontally and 2 cm to 100 pupils vertically.

   (b) Estimate how many pupils score

       (i) less than 16 marks,   (ii) more than 65 marks.

   (c) What percentage of the pupils score more than 65 marks?

8. Sketch the frequency curve corresponding to the cumulative frequency curve shown in Fig. 4.18.

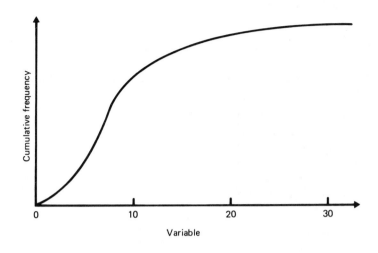

Fig. 4.18.

9. A frequency polygon is shown in Fig. 4.19. Draw the corresponding histogram.

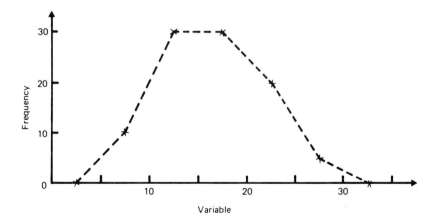

Fig. 4.19.

10. Draw a 'less than' cumulative frequency distribution curve for the data below which relate to the weekly earnings of women aged 18 and over employed in non-manual occupations.

Weekly earnings	Percentage of women
Under £20	3
£20 and under £30	19
£30 and under £40	28
£40 and under £50	22
£50 and under £60	13
£60 and under £70	6
£70 and under £80	4
£80 and under £90	2
£90 and over	3

(a) What percentage earn less than £28 per week?

(b) What percentage earn between £44 and £84 per week?

(c) What percentage earn more than £84 per week?

11. 5 coins were tossed 500 times with the results shown below:

Number of tails	Number of tosses (frequency)
0	19
1	72
2	171
3	143
4	82
5	13
	Total 500

Draw a percentage ogive and use it to estimate the percentage of tosses which resulted in 3 or more tails.

5 STATISTICAL AVERAGES

5.1 INTRODUCTION

In Chapter 4 we saw that a mass of raw data did not mean very much until it was arranged into a frequency distribution or until it was represented by a histogram.

A second way of making the data more understandable is to try to find a single value which will represent all the values in the distribution. This single representative value is called an average.

In statistics several kinds of average are used. The more important are

(i)   the arithmetic mean, often referred to as the mean
(ii) the median
(iii) the mode
(iv) the geometric mean.

5.2 ARITHMETIC MEAN

The arithmetic mean is found by adding up all the values in the set and dividing this sum by the number of values. That is

$$\text{Arithmetic mean} = \frac{\text{sum of all the values}}{\text{the number of values}}$$

EXAMPLE 1. The height of five men are: 177.8 cm, 175.3 cm, 174.8 cm, 179.1 cm and 176.5 cm. Calculate the mean height of the five men.

$$\text{Mean} = \frac{177.8 + 175.3 + 174.8 + 179.1 + 176.5}{5}$$

$$= \frac{883.5}{5} = 176.7 \, \text{cm}$$

5.3 MEAN OF A FREQUENCY DISTRIBUTION

When finding the mean of a frequency distribution we must take into account the frequencies as well as the measured observations.

EXAMPLE 2. 5 packets of chemical have a mass of 20.01 grams, 3 have a mass of 19.98 grams and 2 have a mass of 20.03 grams. What is the mean mass of the packets?

The total mass $= (5 \times 20.01) + (3 \times 19.98 + (2 \times 20.03)$

$$= 100.05 + 59.94 + 40.06 = 200.05 \text{ grams}$$

Total number of packets $= 5 + 3 + 2 = 10$

Mean mass $= \dfrac{200.05}{10} = 20.005 \text{ grams}$

EXAMPLE 3. The mean value of 10 observations is 4.5 but the mean value of 6 of these observations is 5.3. Find the mean value of the remaining 4 observations.

Sum of 10 observations $= 10 \times 4.5 = 45$

Sum of 6 observations $= 6 \times 5.3 = 31.8$

Sum of 4 observations $= 45 - 31.8 = 13.2$

Mean value of the 4 observations $= \dfrac{13.2}{4} = 3.3$

EXAMPLE 4. The mean of three numbers a, b and c is 10 whilst the mean of the five numbers a, b, c, x and y is 16. Calculate the value of x if y is 22.

$a + b + c = 3 \times 10 = 30$

$a + b + c + x + y = 5 \times 16 = 80$

$30 + x + y = 80$

$x + y = 50$

$x + 22 = 50$

$x = 50 - 22 = 28$

Example 5 points the way in which we can find the mean of a frequency distribution.

EXAMPLE 5. Each of 200 similar engine components are measured correct to the nearest millimetre and recorded as follows:

Length (mm)	198	199	200	201	202
Number of components	8	30	132	24	6

Calculate the mean length of the 200 components.

$$\text{Mean length} = \frac{(198 \times 8) + (199 \times 30) + (200 \times 132) + (201 \times 24) + (202 \times 6)}{200}$$

$$= \frac{1584 + 5970 + 26\,400 + 4824 + 1212}{200}$$

$$= \frac{39\,990}{200} = 199.95\,\text{mm}$$

If $x_1, x_2, x_3, \ldots, x_n$ are the measured observations which have frequencies $f_1, f_2, f_3, \ldots f_n$ then the mean of the distribution ($\bar{x}$) is always given by

$$\bar{x} = \frac{x_1 f_1 + x_2 f_2 + x_3 f_3 + \ldots + x_n f_n}{f_1 + f_2 + f_3 + \ldots + f_n} = \frac{\Sigma\, xf}{\Sigma\, f} = \frac{\Sigma\, xf}{n}$$

where n is the total number of observations.

The symbol Σ (Greek letter capital sigma) means the 'sum of'. Thus Σxf tells us to multiply together the corresponding values of x and f and add all these products together. The mean of a frequency distribution is usually found by using a tabular method as shown in Example 6.

EXAMPLE 6. The table below gives the distances travelled during a particular period by 50 lorries belonging to a transport firm. Calculate the mean distance travelled.

Distance (hundreds of km)	48	49	50	51	52	53	54	55
Number of lorries	1	3	7	14	12	8	3	2

x	f	xf
48	1	48
49	3	147
50	7	350
51	14	714
52	12	624
53	8	424
54	3	162
55	2	110
	50	2579

We have: $n = 50$ and $\Sigma\, xf = 2579$

$$\bar{x} = \frac{2579}{50} = 51.58$$

Hence the average distance travelled is $51.58 \times 100 = 5158\,\text{km}$.

5.4   CODED METHOD OF COMPUTING THE MEAN †

A shorter and simpler method of calculating the mean is obtained by using one of the values of x as an assumed mean. The next step is to choose a unit size which is usually the difference between the successive values of x. Thus in Example 6 we can take 50 as our assumed mean and 1 as our unit size. We can now express the values of x in terms of units above or below the assumed mean. Thus when x = 48, d (deviation from the assumed mean) = $48 - 50 = -2$. That is, 48 is 2 units less than the assumed mean. Similarly when x = 54, $d = 54 - 50 = +4$. That is, 54 is 4 units more than the assumed mean. The full set of deviations from the assumed mean are then as follows:

Assumed mean = 50   Unit size = 1

x	48	49	50	51	52	53	54	55
d	−2	−1	0	+1	+2	+3	+4	+5

The calculation of the mean is then as follows:

x	d	f	df
48	−2	1	−2
49	−1	3	−3
50	0	7	0
51	1	14	14
52	2	12	24
53	3	8	24
54	4	3	12
55	5	2	10
		50	79

We have: n = 50 and $\Sigma \, df$ = 79

$$\bar{d} = \frac{\Sigma \, df}{n} = \frac{79}{50} = 1.58$$

$\bar{x}$ = assumed mean + $\bar{d}$ × unit size

$$= 50 + 1.58 \times 1 = 50 + 1.58 = 51.58$$

This is the same as the answer obtained in Example 6 but the arithmetic is a good deal simpler and the calculation may be performed without a calculator.

—————————

† For proof see page 234

The mean of a grouped distribution is found by taking the values of x as the class mid-points as shown in Example 7.

EXAMPLE 7. The table below shows the earnings of men employed by a certain firm. Calculate the mean of the distribution.

Weekly earnings (£)	Number
60 but under 66	2
66 but under 72	8
72 but under 78	14
78 but under 84	23
84 but under 90	12
90 but under 96	6
96 but under 102	2

Assumed mean = 81   Unit size = 6

Class	x	d	f	df
60 – 66	63	−3	2	−6
66 – 72	69	−2	8	−16
72 – 78	75	−1	14	−14
78 – 84	81	0	23	0
84 – 90	87	1	12	12
90 – 96	93	2	6	12
96 – 102	99	3	2	6
			67	−6

We have: $n = 67$ and $\Sigma\, fd = -6$

$$\bar{d} = \frac{-6}{67} = -0.090$$

$$\bar{x} = 81 + (-0.090) \times 6 = 80.46$$

Hence the mean weekly wage is £80.46.

When we have a distribution in which the class widths differ we must first decide upon a standard unit size, as shown in Example 8.

EXAMPLE 8. The table below shows the annual incomes for a group of 34 people. Calculate the mean annual income for this group.

Income	£2000 –	£2500 –	£3000 –	£4000 – 6000
Number of people	2	12	12	8

Assumed mean = £3500   Unit size = £250

Class	x	d	f	fd
2000 – 2500	2250	−5	2	−10
2500 – 3000	2750	−3	12	−36
3000 – 4000	3500	0	12	0
4000 – 6000	5000	6	8	48
			34	2

We have $n = 34$ and $\Sigma fd = 2$

$$\bar{d} = \frac{2}{34} = 0.059$$

$$\bar{x} = 3500 + 0.059 \times 250 = 3500 + 15 = 3515$$

(In this example, having decided on a unit size of £250, £2250 is £1250 less than the assumed mean of £3500, i.e. it is 5 units less than the assumed mean. In the same way £2750 is 3 units less than the assumed mean and £5000 is 6 units more.)

EXERCISE 8

1. Find the mean of £23, £27, £30, £28 and £32.

2. The heights of some men are as follows: 172, 170, 168, 181, 175, 179 and 173 cm. Calculate the mean height of the men.

3. 5 people earn £84 per week, 3 earn £76 per week and 2 earn £88. What is the mean weekly wage of these 10 people?

4. The mean value of 6 observations is 4.2 but 2 of these observations have a mean value of 5. Determine the mean value of the remaining 4 observations.

5. The mean of four numbers a, b, c and d is 8. The mean of the six numbers a, b, c, d, x and y is 7. If the value of x is 8, find the value of y.

6. The four numbers a, b, c and d have a mean value of 6 whilst the five numbers a, b, c, d and e have a mean value of 7. Find the value of e.

7. Calculate the mean length from the following table:

Length (mm)	198	199	200	201	202
Frequency	1	4	17	2	1

8. Estimate the mean of the following data:

Mass (kg)	0 – 4.9	5 – 9.9	10 – 14.9	15 – 19.9	20 – 24.9
Frequency	50	64	43	26	17

9. The marks obtained by a group of 50 students in an examination were as follows:

Marks	10 – 19	20 – 29	30 – 39	40 – 49	50 – 69
No. of students	2	7	31	8	2

Calculate the mean mark obtained.

10. The temperature at noon in a certain city was measured every day for the month of June. The distribution was as follows:

Temperature (°C)	16	17	18	19	20	21	22
Number of days	1	4	6	8	9	1	1

Calculate the mean temperature to the nearest °C.

5.5 WEIGHTED ARITHMETIC MEAN

Sometimes, associated with the numbers $x_1, x_2, x_3, \ldots, x_n$ are certain weighting factors $w_1, w_2, w_3, \ldots, w_n$ depending upon the importance attached to each of the numbers. In such cases the weighted mean is

$$\bar{x} = \frac{w_1 x_1 + w_2 x_2 + w_3 x_3 + \ldots + w_n x_n}{w_1 + w_2 + w_3 + \ldots + w_n} = \frac{\Sigma wx}{\Sigma w}$$

Note the similarity between the formula for a weighted mean and that for the mean of a frequency distribution.

EXAMPLE 9. An examination consists of 3 papers — multiple choice, short answer and standard. The standard paper is weighted twice as much as the other two. If a student obtains marks of 56, 62 and 43 respectively on the multiple choice, short answer and standard papers. Calculate the weighted mean mark obtained.

$$\bar{x} = \frac{56 \times 1 + 62 \times 1 + 43 \times 2}{1 + 1 + 2} = \frac{56 + 62 + 86}{4} = \frac{204}{4} = 51$$

Thus the weighted mean mark obtained by the student is 51.

5.6 GEOMETRIC MEAN

The geometric mean of a set of numbers $x_1, x_2, x_3, \ldots, x_n$ is the nth root of the product of the numbers. Thus

$$G = \sqrt[n]{x_1 x_2 x_3 \ldots x_n}$$

EXAMPLE 10. Find the geometric mean of the numbers 8, 27 and 64.

$$G = \sqrt[3]{8 \times 27 \times 64} = 24$$

In practice, if a scientific calculator is not available, the geometric mean is calculated by using logarithms as shown in Example 11.

EXAMPLE 11. The values of five items are as follows: 29.86, 32.17, 29.75, 31.48 and 28.87. Calculate their geometric mean by using logarithms.

$$G = \sqrt[5]{29.86 \times 32.17 \times 29.75 \times 31.48 \times 28.87}$$

Remembering that to find the logarithm of a product we find the logarithms of each of the numbers and add them together we have

$$\log G = \frac{\log 29.68 + \log 32.17 + \log 29.75 + \log 31.48 + \log 28.87}{5}$$

We divide by 5 because to find the root of a number we divide the logarithm of the number by the number denoting the root, and the antilogarithm giving the required root. The work is best done by using a table as shown below:

Number	Logarithm
29.86	1.4751
32.17	1.5074
29.75	1.4735
31.48	1.4980
28.87	1.4604

By finding the antilogarithm of 1.4829,

$$G = 30.40$$

$$5)\overline{7.4144}$$
$$1.4829$$

5.7 GEOMETRIC MEAN OF A FREQUENCY DISTRIBUTION

If the numbers $x_1, x_2, x_3, \ldots x_k$ occur with the frequencies (or weights $f_1, f_2, f_3 \ldots, f_k$ their geometric mean is

$$G = \sqrt[n]{x_1^{f_1} \cdot x_2^{f_2} \cdot x_3^{f_3} \cdots x_k^{f_k}}$$

where $n = f_1 + f_2 + f_3 +, \cdots, + f_k$

EXAMPLE 12. The table below shows the distribution of maximum loads in tonnes (1 tonne = 1000 kg) supported by certain cables manufactured by a firm. Find the geometric mean of the cables.

Max. load (tonnes)	8.3 – 8.5	8.6 – 8.8	8.9 – 9.1	9.2 – 9.4	9.5 – 9.7
Number of cables	2	8	14	6	1

For a grouped distribution the values of x are taken as the class mid-points. The logarithm of a number raised to a power is the logarithm of the number multiplied by the index denoting the power. Hence we have

$$\log G = \frac{2 \log 8.4 + 8 \log 8.7 + 14 \log 9.0 + 6 \log 9.3 + 1 \log 9.6}{31}$$

The work is best done by using a table as shown below:

Class	x	$\log x$	f	$f \log x$
8.3 – 8.5	8.4	0.9243	2	1.8486
8.6 – 8.8	8.7	0.9395	8	7.5160
8.9 – 9.1	9.0	0.9542	14	13.3588
9.2 – 9.4	9.3	0.9685	6	5.8110
9.5 – 9.7	9.6	0.9823	1	0.9823
			31	31)29.5167
				0.9521

The antilogarithm of 0.9521 is 8.96 and hence $G = 8.96$

The geometric mean of the cables is 8.96 tonnes.

EXERCISE 9

1. The prices of butter, cheese and cooking fat were per kilogram, £1.12, £1.48 and £0.76. If weights of 2, 1 and 3 respectively are to be applied, calculate the weighted mean price of the three foods.

2. A group examination consists of pure maths, applied maths and physics. The physics paper is weighted twice as much as the other two. If a candidate obtains 54 marks for pure maths, 48 marks for applied maths and 43 marks for physics calculate the weighted mean mark obtained.

3. Find the geometric mean of the numbers 2, 4 and 8.

4. Find the geometric mean of the numbers 18.17, 19.48, 18.78 and 17.94.

5. Find the weighted geometric mean of the prices given in Question 1.

6. Calculate the geometric mean of the following frequency distribution:

x	1.23	1.24	1.25	1.26	1.27
Frequency	2	5	12	4	1

7. Calculate the geometric mean of the following frequency distribution which relates to the heights of female workers in a certain factory.

Height (cm)	156 – 158	159 – 161	162 – 164	165 – 167	168 – 170
Number	2	9	18	7	3

5.8 MEDIANS

If a set of numbers is arranged in ascending (or descending) order of size the median is the number which lies half-way along the series. Thus the median of 3, 4, 4, 5, 6, 8, 8, 9, 10 is 6, because there are four numbers below and four numbers above this value.

If there is an even number of values then the median is found by taking the arithmetic mean of the two middle values. Thus the median of 3, 3, 5, 7, 9, 10, 13, 15 is

$$\frac{7+9}{2} = 8$$

5.9 MEDIAN OF A FREQUENCY DISTRIBUTION

The median of an ungrouped discrete frequency distribution can be found by setting out the scores in numerical order and finding the middle value.

EXAMPLE 13. The table below shows the distribution of the numbers obtained when a dice was thrown 30 times. Determine the median.

Number obtained	1	2	3	4	5	6
Frequency	2	7	5	7	3	6

The total frequency is 30 hence the median must lie between the 15th and 16th items in the distribution. We now look at the frequencies to find out the values of the 15th and 16th items. These are both 4 and hence the median number is 4.

It is unnecessary to write down all the items in numerical order to determine the median but if we do this we obtain

1, 1, 2, 2, 2, 2, 2, 2, 2, 3, 3, 3, 3, 3, 4, 4, 4, 4, 4, 4, 4, 5, 5, 5, 6, 6, 6, 6, 6, 6
                                              ↑ ↑

Looking at this set of values we see that the two middle values are 4 and hence the median is 4.

5.10 MEDIAN OF A GROUPED FREQUENCY DISTRIBUTION

When the data are grouped into class intervals it is possible, by looking at the distribution, to determine in which class the median lies. To determine its value more exactly we draw a cumulative frequency curve. The median is then the value of the variable corresponding to half the total frequency. If a percentage ogive is drawn then the median is the value of the variable corresponding to 50% of the total frequency and hence it divides the frequency curve into two equal parts.

EXAMPLE 14. The table below shows the annual rents of people living in council houses in 1972.

Rent (£ per annum)	Under 40	40 –	80 –	120 –	160 –
Percentage of households	2	15	25	27	18

Rent (£ per annum)	200 –	240 – 280	280 and over
Percentage of households	8	3	2

Draw an ogive and from it estimate the median rent.

Annual rent £	Cumulative %
Less than 40	2
Less than 80	17
Less than 120	42
Less than 160	69
Less than 200	87
Less than 240	95
Less than 280	98

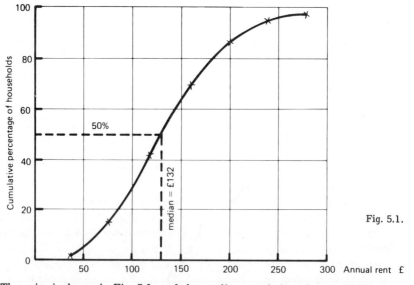

Fig. 5.1.

The ogive is shown in Fig. 5.1, and the median rent is found to be £132.

5.11 MODES

The mode of a set of numbers is the number which occurs most frequently, that is, it is the most common value. Thus the mode of 2, 3, 3, 4, 4, 4, 5, 6, 6, 7 is 4 because it occurs three times which is more times than any of the other numbers in the set.

Sometimes, in a set of numbers, no mode exists, as for instance with the set 2, 4, 7, 8, 9, 11 in which each number occurs once. It is possible for there to be more than one mode. The set 2, 3, 3, 3, 4, 4, 5, 6, 6, 6, 7, 8 has two modes 3 and 6 because each occurs three times which is more than any of the other numbers. A set of values which has two modes is called *bimodal.* If the set has only one mode it is said to be *unimodal,* but if there is more than two modes the set is *multimodal.*

5.12 MODE OF A FREQUENCY DISTRIBUTION

The mode of a frequency distribution is the value of the variable corresponding to the maximum point on the frequency curve (Fig. 5.2). However the mode can also be obtained directly from the histogram as shown in Fig. 5.3. Note that the modal class is the class which contains the mode.

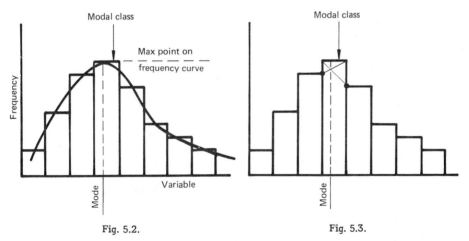

Fig. 5.2.   Fig. 5.3.

EXAMPLE 15. The table below shows the masses of 80 young men measured to the nearest kilogram. Construct a histogram and hence find the mode.

Mass (kg)	40–44	45–49	50–54	55–59	60–64	65–69	70–74	75–79
Number	1	6	14	21	18	12	6	2

The histogram is drawn in Fig. 5.4 and the mode is found to be 58 kg. (Note that the class 55 — 59 is the modal class.)

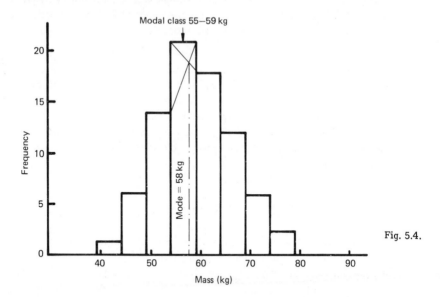

Fig. 5.4.

5.13 COMPARISON OF THE STATISTICAL AVERAGES

| ADVANTAGES | DISADVANTAGES |

The Mean

(a) Is the best known average.

(b) Can be calculated exactly.

(c) Makes use of all the data.

(d) Can be used in further statistical work.

(a) Is greatly affected by extreme values.

(b) When the data are discrete can give an impossible figure (2.364 children).

(c) Cannot be obtained graphically.

The Median

(a) Is not influenced by extreme values.

(b) Can be obtained even if some of the values in a distribution are unknown.

(c) It is unaffected by irregular class widths and is not affected by open-ended classes.

(d) It can represent an actual value in the data.

(a) For grouped distributions its value can only be estimated from an ogive.

(b) When only a few items are available or when the distribution is irregular the median may not be characteristic of the group.

(c) Cannot be used in further statistical calculations.

A First Course in Statistics

ADVANTAGES	DISADVANTAGES

The Mode

(a) Is unaffected by extreme values.

(b) Easy to obtain from a histogram.

(c) To determine its value only values near to the modal class are required.

(a) When the data are grouped its value cannot be determined exactly.

(b) There may be more than one mode.

(c) Cannot be used for further statistical work.

Geometrical Mean

(a) Uses all the available data.

(b) Can be used in further statistical work.

(c) Not so sensitive to extreme values as the mean.

(a) More difficult to calculate than the other averages.

(b) Difficult to understand.

(c) Cannot be calculated if any value is zero or if an odd number of items are negative.

5.14 WHICH AVERAGE TO USE

The arithmetic mean is the most familiar kind of average and it is extensively used in business and other statistical work, such as with sales data, income and expenditure, operation costs, rates of pay, etc. It is easy to understand but in some circumstances it can be definitely misleading. For instance, if the hourly wages of five employees in an office are £1.52, £1.64, £1.88, £4.60 and £1.76. The mean wage is £2.28, but this is affected by the extreme value of £4.60 and hence the value of the mean gives a false impression of the wages paid in the office under discussion.

The median is not affected by extreme values, and it will give a better indication of the wages paid in the office discussed above (note that the median wage is £1.76 per hour). For distributions which are sharply peaked or very skew the median is usually the best average to use.

The mode is used when the commonest value is required. For instance, a manufacturer is not particularly interested in the mean length of men's legs because it might not represent a stock size in trousers. It may, in fact, be some point between stock sizes and in such cases the mode is probably the best average to use. However, which average is used will depend upon particular circumstances.

5.15 RELATION BETWEEN THE MEAN, MEDIAN AND MODE

When the distribution is symmetrical the mean, median and mode will have the same value (Fig. 5.5) but when the distribution is skewed they will have different values (Fig. 5.6).

For frequency distributions which are unimodal and moderately skewed the equation below gives an approximate relationship between the arithmetic mean, the median and the mode:

$$\text{Mean} - \text{mode} = 3(\text{mean} - \text{median})$$

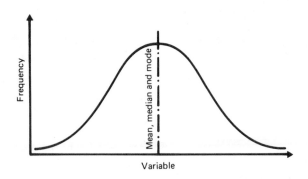

Fig. 5.5. For a symmetrical distribution the arithmetic mean, median and mode have the same value.

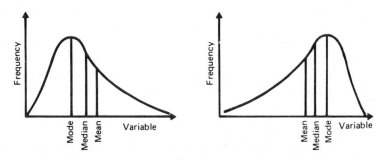

Fig. 5.6. When a frequency distribution is skewed the mean, median and mode have different values.

5.16 QUANTILES

We have seen that the median divides a set of values into two equal parts. Using a similar method we can divide the set into four equal parts. The values which so divide the set are called the *quartiles*. They are often denoted by the symbols Q_1, Q_2, and Q_3, Q_1 being the first or lower quartile, Q_2 being the second quartile and Q_3 the third or upper quartile. The value of Q_2 is equal to the median.

Similarly, the values of the variable which divide a distribution into ten equal parts are called the *deciles* and are denoted by $D_1, D_2, D_3, ..., D_9$. The *percentiles* are the values of the variable which divide the distribution into one hundred equal parts and these are denoted by $P_1, P_2, P_3, ..., P_{99}$.

Collectively, the quartiles, deciles and percentiles are called *quantiles*.

EXAMPLE 16. For the information given in the table below find the upper and lower quartiles, the 6th decile and the 90th percentile.

Number of seats in theatre	Up to 250	251 – 500	501 – 750	751 – 1000
Number of theatres	8	14	25	16

Number of seats in theatre	1001 – 1250	1251 – 1500	more than 1500
Number of theatres	11	5	1

The first step is to obtain a cumulative frequency table:

Number of seats	Cumulative frequency
Less than 250	8
Less than 500	22
Less than 750	47
Less than 1000	63
Less than 1250	74
Less than 1500	79

The next step is to draw the ogive shown in Fig. 5.7.

The lower quartile is the value of the variable corresponding to one-quarter of the total frequency, i.e. ¼ of 80 = 20. Its value, from the ogive is 460. The upper quartile is the number of seats corresponding to three-quarters of the total frequency, i.e. ¾ of 80 = 60. From the ogive the upper quartile is found to be 950 seats.

The 6th decile is the number of seats corresponding to $^6/_{10}$ of the total frequency, i.e. $^6/_{10}$ of 80 = 48. From the ogive the 6th decile is found to be 770 seats. The 90th percentile is the value of the variable corresponding to 90% of the total frequency, i.e. 90% of 80 = 72. The value of the 90th percentile is found to be 1200 seats.

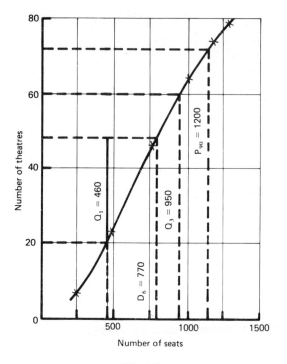

Fig. 5.7.

EXERCISE 10

1. Find the median of the numbers 5, 3, 8, 6, 4, 2, 8.

2. Find the median of the numbers 2, 4, 6, 5, 3, 1, 8, 9.

3. The marks of a student in five examintions were: 54, 63, 49, 78 and 57. Find his median mark.

4. The annual salaries of six office workers are £3580, £4976, £2380, £5900, £8620 and £4320. Determine the median salary.

5. 97 numbers are arranged in ascending order. How would you find the median of the numbers?

6. The table below gives the frequency of the masses 61 − 70, 71 − 80, etc., to the nearest gram, of 40 carrots grown for the frozen food industry.

Mass (g)	61 − 70	71 − 80	81 − 90	91 − 100	101 − 110
Frequency	5	6	19	7	3

(a) Construct a cumulative-frequency table.

(b) Draw the cumulative-frequency curve.

(c) Use your graph to estimate:

    (i) the median,

    (ii) the lower quartile,

    (iii) the upper quartile.

7.   In an experiment in a biology laboratory, the weights of 100 leaves were found. The distribution of weights is given in the following table:

Weight (in grams)	Frequency
0 – 4	1
5 – 9	2
10 – 14	15
15 – 19	21
20 – 24	26
25 – 29	28
30 – 34	4
35 – 39	2
40 – 44	1

(a) Draw up a table to show the cumulative frequencies.

(b) Draw a cumulative frequency graph (ogive) and from your graph determine the upper and lower quartile.

8.   The table below shows the number of calls per day received by a fire station over a given year. Determine the median of this distribution.

No. of calls per day	0	1	2	3	4	5	6 and over
No. of days	139	102	57	30	19	12	6

9. The table below gives the frequency distribution of marks obtained by 500 candidates in a statistics examination.

Mark	0 – 9	10 – 19	20 – 29	30 – 39	40 – 49
Frequency of marks	5	10	45	65	105

Mark	50 – 59	60 – 69	70 – 79	80 – 89	90 – 99
Frequency of marks	120	75	40	25	10

Construct a table showing the cumulative frequency distribution and draw the cumulative frequency curve (i.e. the ogive). Estimate from your graph:

(a) the median mark

(b) the mark at the 90th percentile

(c) the number of candidates who got less than 25 marks

(d) the percentile for a mark of 40.

Briefly explain how you obtained each result. Check, by calculation and without reference to your graph, the accuracy of your four estimates. Comment on any assumptions you make.

10. (a) Draw a sketch-graph of:

   (i) a normal distribution curve

   (ii) a postively skewed distribution curve.

In the case of each of (i) and (ii) indicate on your diagram the estimated position of the mean of the distribution.

   (b) Fifteen pupils are asked to estimate the length (to the nearest 2 cm) of their teacher's table. The estimates, arranged in descending order, were as follows:

148 146 144 142 140 140 138 138 138 136 136 134 132 132 128

Find the median estimate and the lower and upper quartiles.

11. Thirteen people were asked to guess the mass of a cake to the nearest half kilogram. The results were:

3½, 2½, 2, 1, 3½, 2, 3½, 3, 3, 1, 1½, 2½, 3½ kg.

What was:
(a) the modal value
(b) the median value
(c) the arithmetic mean
(d) the lower quartile
(e) the upper quartile.

12. The table below shows the frequency distribution of the life of 200 electric light bulbs, which were tested by the firm making them.

Life in hours	Number of bulbs
301 – 400	7
401 – 500	23
501 – 600	28
601 – 700	36
701 – 800	37
801 – 900	28
901 – 1000	22
1001 – 1100	16
1101 – 1200	3

(a) State:

(i) the class interval size
(ii) the modal class.

(b) Calculate the percentage of light bulbs having a life greater than 900 hours.

(c) Using a scale of 1 cm to represent 100 hours and also 10 light bulbs, construct a cumulative frequency curve.

(d) Use the graph to determine the median life of the electric light bulbs.

13. The following figures are the maximum temperatures recorded in degrees Centigrade on 25 consecutive days at a local ice rink.

2	5	3	2	4
5	3	7	6	7
6	7	4	2	5
7	6	3	7	3
3	4	7	5	5

(a) Complete the following frequency table.

Maximum temperature (°C)	2	3	4	5	6	7
Frequency (number of days)						

(b) Draw a histogram to illustrate these statistics.

(c) (i) What is the modal temperature?

(ii) Why does the modal temperature give a misleading impression of the average temperature?

(d) Find the median temperature by inspection.

14. The numbers of peas per pod were counted for a number of pods and the results are tabulated below. Determine the mode.

No. of peas per pod	5	6	7	8	9	10	11
No. of pods	12	25	18	16	10	6	2

15. The hand spans of 40 children were measured to the nearest cm. The results are printed below.

15	19	20	16	18	21	23	17	17	18
19	17	18	20	16	13	22	19	18	18
18	19	19	20	17	18	20	22	19	17
18	17	17	16	19	25	15	16	19	18

(a) Copy and complete the frequency table of the distribution.

Hand span to nearest cm	13	14	15	16	17	18	19	20	21	22	23	24	25
Number of children	1	0	2								1	0	1

(b) Draw a histogram to represent the results.

(c) What is the mode of the distribution?

16. The table below shows the distribution of the scores obtained by 200 pupils in a particular test.

Score	No. of pupils
76 – 80	2
71 – 75	5
66 – 70	9
61 – 65	19
56 – 60	32
51 – 55	55
46 – 50	38
41 – 45	21
36 – 40	10
31 – 35	6
26 – 30	3

Construct a cumulative frequency curve and use it to estimate

(a) the median score

(b) the value of the lower quartile

(c) the number of students scoring more than 42

(d) the 60th percentile

(e) the percentile corresponding to a score of 68.

17. Sketch a frequency distribution in which the mode is larger than the mean.

18. An ordinary die was thrown a number of times and the scores obtained, with the exception of the number of scores of 1, are shown below.

Score	1	2	3	4	5	6
Frequency		5	2	5	3	2

If the median of the complete distribution is 2.5, how many scores of 1 must have been thrown?

19. A variable x has a moderately skew distribution as in Fig. 5.8. The vertical at A divides the area under the curve into two equal parts. State the usual name for the value A.

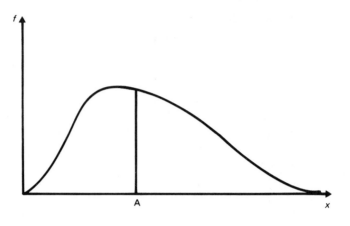

Fig. 5.8.

20. The table shows the number of tests required to pass the driving examination for the pupils of a particular driving school.

Number of tests required to pass	1	2	3	4	5
Men	5	11	8	7	3
Women	7	5	3	0	1

(a) State the mode of the number of tests required to pass the examination for men drivers.

(b) State the median of the number of tests required for women drivers.

6 MEASURES OF DISPERSION

6.1 INTRODUCTION

A statistical average gives some idea about the position of a distribution (Fig. 6.1) and hence statistical averages are often called measures of location. We now need measures which will define the spread or dispersion of the data. The measures of dispersion which are most often used are:

(i)   the *range*

(ii)   the *mean deviation*

(iii) the *quartile deviation* which is sometimes called the semi-interquartile range

(iv) the *standard deviation*.

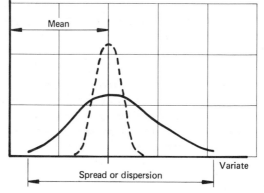

Two different distributions with the same mean

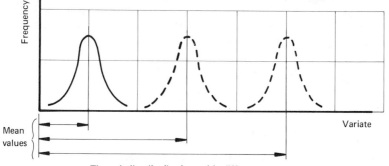

Three similar distributions with different means

Fig. 6.1.

97

6.2 RANGE

The range is the difference between the largest and smallest observations in a distribution. That is,

range = largest observation − smallest observation

EXAMPLE 1. The wages paid in an office are £57, £62, £84, £68.40 and £80 per week. Find the range of these wages.

> lowest wage = £57
>
> highest wage = £84
>
> range of wages = £84 − £57 = £27

EXAMPLE 2. Find the range of loads for the distribution shown in the table below.

Max. load (tonnes)	8.3 − 8.5	8.6 − 8.8	8.9 − 9.1	9.2 − 9.4
Frequency	2	8	6	1

The lower boundary of the first class is 8.25 tonnes and the upper boundary of the last class is 9.45 tonnes. Hence

> range of max. loads = 9.45 − 8.25 = 1.2 tonnes

The range gives some idea of the spread of a distribution but it depends solely upon the extreme values of the data. It gives no information about the way in which the data are dispersed and hence it is seldom used as a measure for a frequency distribution. However, when dealing with small samples the range is a very effective measure of dispersion.

6.3 MEAN DEVIATION

The mean deviation is the mean of the deviations of the data from a statistical average which may be the mean, median or mode. The deviation is the difference between the value of the average and a value in the distribution. The mean deviation is calculated from the formula

$$\text{mean deviation} = \frac{\Sigma |x - \bar{x}|}{n}$$

where x = a value in the distribution

$\bar{x}$ = a statistical average

and n = the number of items in the distribution.

The lines || indicate that all the differences are to be taken as positive.

EXAMPLE 3. Find the mean deviation from the mean for the numbers 3, 5, 7, 9 and 11.

$$\bar{x} = \frac{3+5+7+9+11}{5} = \frac{35}{5} = 7$$

x	$|x - \bar{x}|$
3	4
5	2
7	0
9	2
11	4
$\Sigma|x - \bar{x}| = 12$	

Since there are five numbers, $n = 5$. Hence

$$\text{mean deviation} = \frac{12}{5} = 2.4$$

EXAMPLE 4. Find the mean deviation from the median of the numbers 8, 2, 11, 5 and 9. Arranging the numbers in ascending order:

2, 5, 8, 9, 11

we see that the median is 8.

x	$x - \bar{x}$
2	6
5	3
8	0
9	1
11	3
$\Sigma|x - \bar{x}| = 13$	

$$\text{mean deviation} = \frac{13}{5} = 2.6$$

Measures of Dispersion

Note that the signs of the deviations from the statistical average are ignored. In the case of the mean deviation from the mean (as in Example 3) the sum of the deviations would be zero if the signs of the deviations were not ignored.

6.4 MEAN DEVIATION OF A FREQUENCY DISTRIBUTION

The mean deviation of a frequency distribution is found by using the following formula:

$$\text{mean deviation} = \frac{\Sigma f|x - \bar{x}|}{n}$$

where $\bar{x}$ = the value of the statistical average used
  x = the value of the class mid-point
  f = the corresponding class frequency
  n = the number of items in the distribution = Σf

EXAMPLE 5. The table below gives the distribution of the heights of 100 men. Calculate the mean deviation of these heights from the mean.

Height (cm)	150 – 154	155 – 159	160 – 164	165 – 169	170 – 174
Frequency	8	16	43	29	4

Class	x	f	fx	$|x - \bar{x}|$	$f|x - \bar{x}|$
150 – 154	152	8	1216	10.25	82.00
155 – 159	157	16	2512	5.25	84.00
160 – 164	162	43	6966	0.25	10.75
165 – 169	167	29	4843	4.75	137.75
170 – 174	172	4	688	9.75	39.00
		100	16225		353.50

$$\bar{x} = \frac{\Sigma fx}{n} = \frac{16\,225}{100} = 162.25$$

$$\text{mean deviation from the mean} = \frac{\Sigma f|x - \bar{x}|}{n} = \frac{353.50}{100} = 3.535 \text{ cm}$$

The mean deviation is used in economic and social statistics. The lower its value, the smaller is the spread of the data.

6.5 QUARTILE DEVIATION

The quartile deviation, sometimes called the semi-interquartile range, is found by using the formula:

quartile deviation $= \frac{1}{2}(Q_3 - Q_1)$

where Q_3 is the upper quartile and Q_1 is the lower quartile.

EXAMPLE 6. An examination of the wages paid by a certain company showed that the upper quartile was £84 per week whilst the lowest quartile was £68 per week. Find the semi-interquartile range.

We are given that Q_3 = £84 and that Q_1 = £68 per week. Hence

Semi-interquartile range $= \frac{1}{2}(84 - 68) = \frac{1}{2} \times 16 = £8$ per week

For a frequency distribution the values of the upper and lower quartiles may be found by drawing a cumulative frequency curve (see pages 89–91) and hence the quartile deviation may be found.

The quartile deviation has the advantage that it ignores extreme values of a distribution. However, this measure of dispersion only covers that half of the distribution which is centred upon the median and therefore it does not show the dispersion of the distribution as a whole. A small value of the quartile deviation shows that the data has only a small amount of spread between the quartiles. The quartile deviation is used extensively in business and educational statistics.

6.6 STANDARD DEVIATION

The standard deviation is the most important of the measures of dispersion. It is usually denoted by the symbol σ (Greek letter small sigma) and it may be calculated by using either of the following formulae:

$$\sigma = \sqrt{\frac{\Sigma(x - \bar{x})^2}{n}} \quad \text{or} \quad \sigma = \sqrt{\frac{\Sigma x^2}{n} - (\bar{x})^2}$$

EXAMPLE 7. Find the standard deviation of the numbers 2, 4, 7, 8 and 9.

(i)   Using the formula $\sigma = \sqrt{\dfrac{\Sigma(x - \bar{x})^2}{n}}$

$$\bar{x} = \frac{2 + 4 + 7 + 8 + 9}{5} = 6$$

x	$x - \bar{x}$	$(x - \bar{x})^2$
2	-4	16
4	-2	4
7	1	1
8	2	4
9	3	9
$\Sigma(x - \bar{x})^2$		34

$$\sigma = \sqrt{\frac{\Sigma(x - \bar{x})^2}{n}}$$

$$= \sqrt{\frac{34}{5}} = \sqrt{6.8} = 2.61$$

(ii)   Using the formula $\sigma = \sqrt{\dfrac{\Sigma x^2}{n} - (\bar{x})^2}$

x	x^2
2	4
4	16
7	49
8	64
9	81
$\Sigma x^2 = 214$	

$$\sigma = \sqrt{\frac{\Sigma x^2}{n} - (\bar{x})^2} = \sqrt{\frac{214}{5} - 6^2}$$

$$= \sqrt{42.8 - 36} = \sqrt{6.8} = 2.61$$

6.7   STANDARD DEVIATION OF A FREQUENCY DISTRIBUTION

The standard deviation of a frequency distribution is easiest calculated by using a coded method and an assumed mean (see page 76). When this method is used the coded value of the standard deviation is found by using the formula:

$$\sigma_c = \sqrt{\frac{\Sigma f d^2}{n} - \bar{d}^2} \; \dagger \qquad \sigma = \sigma_c \times \text{unit size}$$

where d = the deviation from the assumed mean

$\bar{d}$ = the arithmetic mean of the deviations from the assumed mean

f = the class frequency

n = the number of items in the distributions = Σf

σ_c = the coded value of the standard deviation.

† For proof see page 235

A First Course in Statistics

EXAMPLE 8

(a) The table which follows shows the distribution of maximum loads supported by certain chains. Calculate the mean and standard deviation for this information.

Max load (tonnes)	8.3 – 8.5	8.6 – 8.8	8.9 – 9.1	9.2 – 9.4	9.5 – 9.7
Number of chains	2	8	14	6	1

(b) A further sample of 19 chains was checked and this sample gave a mean load of 9.032 tonnes with a standard deviation of 0.300 tonnes. Calculate the mean and standard deviation of the 50 chains.

(a)

Class	x	d	f	fd	fd^2
8.3 – 8.5	8.4	−2	2	−4	8
8.6 – 8.8	8.7	−1	8	−8	8
8.9 – 9.1	9.0	0	14	0	0
9.2 – 9.4	9.3	1	6	6	6
9.5 – 9.7	9.6	2	1	2	4
			31	−4	26

Assumed mean = 9.0 tonnes
Unit size = 0.3 tonnes

$$\bar{d} = \frac{\Sigma fd}{n} = \frac{-4}{31} = -0.13$$

$\bar{x}$ = assumed mean $\pm \bar{d}$ × unit size (see page 76).

$$= 9.0 + (-0.13) \times 0.3 = 9.0 - 0.039 = 8.961$$

Hence the mean of the distribution is 8.961 tonnes.

$$\sigma_c = \sqrt{\frac{\Sigma fd^2}{n} - (\bar{d})^2}$$

$$= \sqrt{\frac{26}{31} - (-0.13)^2} = \sqrt{0.839 - 0.017} = \sqrt{0.822} = 0.907$$

$\sigma = \sigma_c$ × unit size = $0.907 \times 0.3 = 0.272$

Hence the standard deviation of the distribution is 0.272 tonnes.

A rough check on the calculated value of the standard deviation may be made by dividing the range of these data by six. Thus

$$\text{rough check for } \sigma = \frac{9.75 - 8.25}{6} = 0.25$$

This rough check compares fairly well with the value of 0.272 and we know, from the rough check, that the calculated value of the standard deviation is feasible.

(b) $\bar{x} = \dfrac{\Sigma fx}{n}$ or $\Sigma fx = n\bar{x}$

For the distribution of 31 chains we have $\bar{x} = 8.961$.

$\therefore fx = 31 \times 8.961 = 277.8$

For the additional sample of 19 chains we have $\bar{x} = 9.032$

$\therefore fx = 19 \times 9.032 = 171.6$

Hence for the 50 chains

$$\bar{x} = \frac{277.8 + 171.6}{50} = 8.988$$

Hence the mean load for the 50 chains is 8.988 tonnes.

$$\sigma = \sqrt{\frac{\Sigma f(x - \bar{x})^2}{n}} \text{ or } f(x - \bar{x})^2 = n\sigma^2$$

For the distribution of 31 chains we have $\sigma = 0.272$

$\therefore \Sigma f(x - \bar{x})^2 = 31 \times (0.272)^2 = 2.294$

For the additional sample of 19 chains we have $= 0.300$

$\therefore \Sigma f(x - \bar{x})^2 = 19 \times (0.300)^2 = 1.710$

Hence for the 50 chains the standard deviation is

$$\sigma = \sqrt{\frac{2.294 + 1.710}{50}} = 0.283$$

Therefore the standard deviation for the 50 chains is 0.283 tonnes.

The standard deviation is the most important of the measures of dispersion. Every value in the distribution is used in its calculation and it is expressed in the same units as the original data and the arithmetic mean which is always associated with it. The larger the spread of the data the larger is the value of the standard deviation.

The standard deviation is used extensively in more advanced statistical work. It has the disadvantage, however, that it is more difficult to determine than the other measures of dispersion and it also gives undue weight to extreme values in the distribution because deviations from the mean are squared.

6.8   EFFECT OF ADDING A CONSTANT AMOUNT TO EACH VARIABLE

If a constant amount is added to (or subtracted from) each variable in a set of numbers, the mean is increased (or decreased) by the same amount but the standard deviation remains unaltered†.

EXAMPLE 9. The mean of the numbers $x_1, x_2, \ldots, x_n$ is 8 and the standard deviation is 2. Find the mean and standard deviation of the numbers $x_1 - 3, x_2 - 3, \ldots, x_n - 3$.

Each number in the set is decreased by 3 and hence the new mean is $8 - 3 = 5$. The standard deviation remains unaltered at 2.

6.9   EFFECT OF MULTIPLYING EACH VARIABLE BY THE SAME AMOUNT

If each variable in a set is multiplied by the same amount then the mean and standard deviation of these variables is multiplied by the same amount ††.

EXAMPLE 10. The mean of the numbers $x_1, x_2, \ldots, x_n$ is 7 and their standard deviation is 1.5. Find the mean and standard deviation of the numbers $4x_1$, $4x_2 \ldots, 4x_n$. Each number in the set has been multiplied by 4. Hence the new mean is $4 \times 7 = 28$ and the new standard deviation is $4 \times 1.5 = 6$.

6.10 VARIANCE

The variance of a distribution is the square of the standard deviation. Thus

variance $= \sigma^2$

EXAMPLE 11. The standard deviation of a distribution is 1.85. Determine the variance.

Since $\sigma = 1.85$, the variance $= 1.85^2 = 3.42$

6.11 STANDARDISED SCORES

Measures from different distributions can be compared by using standardised scores. A score in a distribution can be expressed in terms of the mean and standard deviation by using the formula

$$z = \frac{x - \bar{x}}{\sigma}$$

† For proof see page 235
†† For proof see page 236

EXAMPLE 12. A student obtained 84 marks in a mathematics examination for which the mean mark was 70 with a standard deviation of 10. Calculate the standardised score for this student.

We are given $x = 84, \bar{x} = 70$ and $\sigma = 10$. Hence

$$z = \frac{x - \bar{x}}{\sigma} = \frac{84 - 70}{10} = 1.4$$

EXAMPLE 13. A girl obtains 85 marks in a final examination in biology for which the mean mark was 65 with a standard deviation of 8 marks. In history she obtained 68 marks but in this subject the mean mark was 54 with a standard deviation of 7 marks. In which examination did she do best?

In biology: $x = 85, \bar{x} = 65$ and $\sigma = 8$, hence $z = \dfrac{85 - 65}{8} = 2.5$

In history: $x = 68, \bar{x} = 54$ and $\sigma = 7$, hence $z = \dfrac{68 - 54}{7} = 2.0$

It can be seen that the girl had a higher standardised score in biology and therefore she has done better in biology than in history.

6.12 CONVERSION TO AN ARBITRARY SCALE

Any score or set of marks can be adjusted to an arbitrary scale with a given mean and standard deviation by using the formula

$$t = \bar{x}_A + \frac{\sigma_A}{\sigma} (x - \bar{x})$$

where t = the score in the arbitrary scale

σ_A = standard deviation of the arbitrary scale

σ = standard deviation of the original scale

$\bar{x}_A$ = arithmetic mean of the arbitrary scale

$\bar{x}$ = arithmetic mean of the original scale

x = the score in the original scale

EXAMPLE 14. A student scored 78 marks in a physics test. The set of marks is to be adjusted to an arbitrary scale which has a mean of 50 and a standard deviation of 20. What is the final mark of the student if the original mean mark was 60 with a standard deviation of 30?

A First Course in Statistics

We are given that $x = 78, \bar{x}_A = 50, \sigma_A = 20, \bar{x} = 60$ and $\sigma = 30$. Hence

$$t = 50 + \frac{20}{30} \times (78 - 60) = 50 + 12 = 62$$

Hence the final mark in the arbitrary scale is 62.

EXAMPLE 15. An examination in physics with chemistry consisted of two papers, one in physics and the other in chemistry. An analysis of the distributions for the two sets of marks gave the following data:

	Max. possible mark	Arithmetic mean	Standard deviation
Physics	120	60	30
Chemistry	105	45	15

In order to give equal importance to the two subjects each set of marks was adjusted to an arbitrary scale with an arithmetic mean of 50 and a standard deviation of 20 before adding them together to produce the final set of marks. Calculate:

(a) the final mark of a student who scored 75 marks in physics and 15 marks in chemistry

(b) the maximum possible score in the final set of physics with chemistry marks

(a) The final mark in physics is

$$t_P = \bar{x}_A + \frac{\sigma_A}{\sigma_P} (x_P - \bar{x}_P)$$

Since $\bar{x}_A = 50, \sigma_A = 20, \sigma_P = 30, \bar{x}_P = 60$ and $x_P = 75$

$$t_P = 50 + \frac{20}{30} \times (75 - 60) = 50 + 10 = 60$$

The final mark in chemistry is

$$t_C = \bar{x}_C + \frac{\sigma_A}{\sigma_C} (x_C - \bar{x}_C)$$

Since $\bar{x}_A = 50, \sigma_A = 20, \sigma_C = 15, \bar{x}_C = 45$ and $\bar{x}_C = 15$,

$$t_C = 50 + \frac{20}{15} \times (15 - 45) = 50 - 40 = 10$$

The final mark in the physics with chemistry examination is

$$t = t_P + t_C = 60 + 10 = 70$$

(b) The maxiumum score in physics is

$$t_P = 50 + \frac{20}{30} \times (120 - 60) = 50 + 40 = 90$$

The maximum score in chemistry is

$$t_C = 50 + \frac{20}{15} \times (105 - 45) = 50 + 80 = 130$$

The maximum possible score is $90 + 130 = 220$.

6.13 SKEWNESS

It was shown in Chapter 5 that for a symmetrical distribution (i.e. a normal distribution) the arithmetic mean, the median and the mode all have the same value. However if the distribution is skewed the mean, median and mode all have different values. (Fig. 5.6).

There are several ways in which the degree of skewness can be measured but the one which is generally used is *Pearson's coefficient of skewness* which is calculated from the formula

$$\text{skewness} = \frac{3 \times (\text{mean} - \text{median})}{\text{standard deviation}}$$

This formula automatically gives the direction of the skew. If the result is positive the skew is positive; if the result is negative the skew is negative. It will be remembered that for positive skew the longer tail is to the right and for negative skew the longer tail is to the left.

The higher the value of Pearson's coefficient the greater the degree of skewness. For a symmetrical distribution the coefficient is zero but it can take any value between -3 and $+3$.

EXERCISE 11

1. The wages of five office workers are £59.50, £55.60, £98.40, £64.80 and £74.80 per week. Determine the range of the wages.

2. The largest of 50 measurements is 29.88 cm. If the range is 0.12 cm, find the smallest measurement.

3. Find the range of the following discrete distribution:

Number of children in a family	0	1	2	3	4	5	6
Number of families	8	16	40	32	20	12	2

4. Find the mean deviation from the arithmetic mean of the numbers 3, 5, 7 and 9.

5. Find the mean deviation from the median for the numbers 2.3, 3.4, 2.8, 4.1 and 2.9

6. Calculate the mean deviation from the arithmetic mean for the following distribution:

Length (mm)	167	168	169	170	171
Frequency	2	8	15	6	3

7. The table below shows the age distribution of the UK at the Census in a certain year.

    (a) Copy and complete the cumulative frequency table below, assuming a maximum age of 100 years.

Age Group	Population (millions)	Cumulative frequency
Under 10	9	
10 – 20	8	
20 – 30	7	
30 – 40	7	
40 – 50	7	
50 – 60	7	
60 – 70	5	
70 – 80	3	
80 and over	1	

    (b) Draw a cumulative frequency graph.

    (c) From the graph determine the values of the lower quartile, the median and the upper quartile.

    (d) Write down the value of the semi-interquartile range.

8. Eleven people were asked to guess the mass of a cake to the nearest half kilogram. The results were:

    2½, 1½, 2, 2½, 1½, 2½, 1, 1½, 2½, 2½, 1 kg.

    State: (a) the modal value   (d) the lower quartile

          (b) the median value   (e) the upper quartile

          (c) the full range   (f) the semi-interquartile range.

9. In an investigation into the use of telephone facilities by a company, the number of calls per day was as follows:

Number of calls	Number of days
250 and less than 280	16
280 and less than 310	27
310 and less than 340	68
340 and less than 370	35
370 and less than 400	14

(a) Draw an ogive for this information.

(b) From the ogive obtain the values of the lower and upper quartiles.

(c) State the value of the quartile deviation.

10. Calculate the mean and standard deviation of the numbers 2, 3, 6, 1, 5 and 7.

11. Calculate the mean and standard deviation of the numbers 8, 11, 12, 14, 15 and 18. Hence, or otherwise, write down the mean and standard deviation of the numbers:

(a) 80, 110, 120, 140, 150 and 180.

(b) 100, 130, 140, 160, 170 and 200.

12. Calculate the mean and standard deviation of the following distribution:

Length (cm)	15	16	17	18	19	20
Frequency	1	6	12	15	7	2

13. The diameters of 200 ball bearings were measured with the following results:

Diameter (mm)	5.94 – 5.96	5.97 – 5.99	6.00 – 6.02	6.03 – 6.05	6.06 – 6.08
Frequency	8	37	90	52	13

Calculate the mean and standard deviation.

14. (a) Calculate the mean and standard deviation for the following distribution:

Height (cm)	Frequency
153 – 157	4
158 – 162	11
163 – 167	20
168 – 172	24
173 – 177	17
178 – 182	4

(b) 120 further heights were measured and gave a mean of 169.0 cm with a standard deviation of 6.204 cm. Calculate the mean and standard deviation for the combined sample of 200 heights.

15. The mean and standard deviation for the numbers $x_1, x_2, \ldots, x_n$ is 7 and 2 respectively. Write down the mean and standard deviation for the numbers:

(a) $3x_1, 3x_2, \ldots, 3x_n$.

(b) $x_1 + 5, x_2 + 5, \ldots, x_n + 5$.

16. A candidate in an examination obtained 58 marks. The mean mark for all the candidates who sat the examination was 52 marks with a standard deviation of 20. Calculate the standardised score for the candidate.

17. A students marks for English and French were 48 and 60 respectively. In English the mean mark was 50 with a standard deviation of 20 and in French the mean mark was 62 with a standard deviation of 15. In which subject did the student do best?

18. An examination in mathematics was held. An analysis of the distributions for the two sets of marks gave the following data:

Max. possible mark   120

Mean mark   58

Standard deviation   16

The set of marks was adjusted to an arbitrary scale with a mean of 50 and a standard deviation of 20.

(a) What is the maximum possible mark in the arbitrary scale?

(b) Calculate the score in the arbitrary scale of a candidate who scored 70 marks in the examination.

19. A class of students is given a test in arithmetic, history and reading. An analysis of the data gave the following results:

Subject	Mean mark	Standard deviation
Arithmetic	62	5
History	76	7
Reading	60	6

In order to give equal importance to the three tests each set of marks was adjusted to an arbitrary scale with a mean mark of 50 and a standard deviation of 14. Calculate the scores in the arbitrary scale of a student who scored 67 in arithmetic, 82 in history and 72 in reading.

20. An examination consists of two papers, Paper 1 and Paper 2. An analysis of the distributions for the two sets of marks gave the following data:

Paper	Arithmetic mean	Standard deviation
1	65	24
2	48	18

In order to give equal importance to the two papers each set of marks was adjusted to an arbitrary scale with a mean of 50 and a standard deviation of 20 before adding them together to produce the final set of marks. Calculate the final mark of a student who scored 72 marks in Paper 1 and 32 marks in Paper 2.

21. The standard deviation of the set of numbers $x_1, x_2, \ldots, x_n$ is 3.

(a) What is the variance?

(b) Calculate the variance of the set of numbers $2x_1, 2x_2, \ldots, 2x_n$.

22. For a skewed distribution the mean is 22, the median is 20 and the standard deviation is 4. Calculate Pearson's coefficient of skewness and sketch the distribution curve.

23. A distribution has a mean of 35 and a median of 40, the standard deviation being 5. Calculate Pearson's coefficient of skewness and make a sketch of the distribution curve.

24. The table below gives the distribution of the numbers of examination passes obtained by 30 pupils in a particular class.

Number of passes	3	4	5	6	7	8
Boys	1	2	4	2	1	2
Girls	3	2	2	6	4	1

(a) Show that the mean deviation from the median of the distribution of passes for the boys is the same as that for the girls.

(b) Calculate the standard deviation of the distribution for the whole class.

25. The following table shows the weights in kilograms, of 250 boys; each weight was recorded to the nearest 100 grams.

Weight (kg)	44.0 – 47.9	48.0 – 51.9	52.0 – 55.9	56.0 – 57.9
No. of boys	3	17	50	45

Weight (kg)	58.0 – 59.9	60.0 – 63.9	64.0 – 67.9	68.0 – 71.9
No. of boys	46	57	23	9

Draw the ogive and use it to estimate:

(a) the semi-interquartile range,

(b) the second decile,

(c) the eighty-fourth percentile,

(d) the percentage of boys weighing over 59 kilograms.

26. 32 children in a class were asked to estimate the length of a metal rod to the nearest cm, and the table below shows the results obtained.

Estimated length (cm)	Frequency
35	1
36	3
37	4
38	8
39	6
40	5
41	3
42	2

(a) What is the range of these estimates?

(b) Draw up a cumulative frequency table and plot a cumulative frequency curve from this information. Take 2 cm to represent 1 cm on the length axis and 2 cm to represent 5 units on the cumulative frequency axis.

(c) From the curve find the median and the semi-interquartile range of the estimated lengths.

27. The table shows the marks a class of 15 children obtained in mathematics and English examinations.

The two marks were added together in order to obtain class positions.

(a) Find the lower quartile, median, and upper quartile marks for mathematics.

(b) Find the lower quartile, median, and upper quartile marks for English.

(c) Find the semi-interquartile range of

    (i) the mathematics marks

    (ii) the English marks.

(d) Find the median of the total marks.

	Maths	English	Total	Position
A	90	40	130	4
B	80	54	134	1
C	72	60	132	2
D	70	46	116	6
E	65	66	131	3
F	56	39	95	5
G	40	47	87	8
H	32	51	83	9
I	30	50	80	12
J	27	38	65	14
K	21	70	91	7
L	20	61	81	10 =
M	17	39	56	15
N	16	65	81	10 =
0	4	67	71	13

28. For the numbers 6, 8, 10, 12, 14, calculate:

    (a) the mean,

    (b) the variance,

    (c) the standard deviation.

29. The masses of 30 ten-week-old gerbils are summarised below.

Mass (g)	38 –	40 –	42 –	44 –	48 –
Number of gerbils	13	9	5	3	0

(a) Write down the mid-value of each of the first four classes.

(b) Estimate the arithmetic mean and the standard deviation of the masses, giving your answers to three significant figures.

30. Number of hours of part-time work in 15 weeks:

    5, 8, 8, 11, 11, 11, 12, 13, 14, 15, 15, 16, 18, 20, 21.

State (a) the median

    (b) the upper quartile,

    (c) the lower quartile,

    (d) the full range,

    (e) the semi-interquartile range.

MISCELLANEOUS EXERCISE

EXERCISE 12

(All questions are of the type usually found in O-level, CSE and similar examination papers.)

SHORT ANSWER QUESTIONS

1. A car travels 10 000 metres in 100 seconds. If both the distance and the time are accurate to 1%, between what limits does the average speed lie?

2. The heights of 1500 men are distributed with a mean height of 175 cm and a variance of 25 cm^2. State the values between which most of the heights are likely to fall.

3. The masses of 200 men were measured to the nearest kilogram. The results are shown in the table below. Estimate the median mass.

Mass (kg)	45 – 54	55 – 64	65 – 74	75 – 84	85 – 94	95 – 104
Frequency	24	50	58	35	21	12

4. If the standard deviation of the numbers $x_1, x_2, \ldots, x_n$ is 4, find the standard deviation of the numbers

    (a) $x_1 - 2, x_2 - 2, \ldots, x_n - 2$

    (b) $4x_1 \ 4x_2, \ldots, 4x_n$

5. Three consecutive classes of a grouped frequency distribution have the following mid-points: 8, 10 and 13. The first two classes are of the same size. Find the limits of the third class.

6. Write down one advantage which the geometric mean has over the arithmetic mean as a measure of location.

7.

Income(£)	3000 –	3500 –	4000 –	5000 – 7000
No. of persons	1000	600	600	400

Fig. 1. shows part of the histogram of the information on the previous page. Draw the complete histogram.

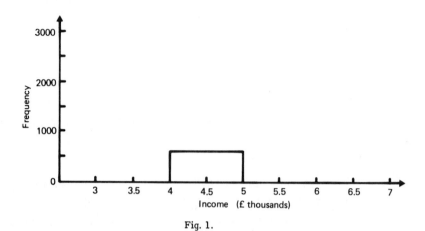

Fig. 1.

8.   Why is the mean deviation used less often than the standard deviation as a measure of dispersion?

9.   A group of ten observations has an arithmetic mean of 10.2 and a mean deviation from the arithmetic mean of 1.3. When two further observations are taken the arithmetic mean and the mean deviation from the arithmetic mean of the twelve observations are the same as before. State the values of these two observations.

10. Draw a frequency distribution in which the mode is larger than the arithmetic mean.

11. What kind of information would be best depicted by a sectional bar chart like the one shown in Fig. 2.

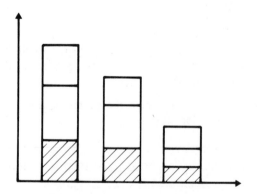

Fig. 2.

12. Find the range of the following set of values:

3, 6, 13, 10, 2, 8, 11, 5, 4 and 12

13. 100 similar parts were measured correct to the nearest millimetre. The results were as follows:

Length (mm)	99	100	101	102	103
Frequency	4	15	66	12	3

Estimate the percentage of parts with a length greater than 102 mm.

14. Describe a situation in which the median is a more useful average than the arithmetic mean.

15. Sketch an ogive corresponding to the frequency distribution shown in Fig. 3.

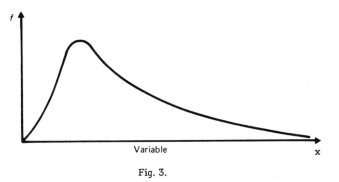

Fig. 3.

16. Determine the range of the following distribution.

No. of peas in a pod	1	2	3	4	5	6	7	8
No. of pods	2	5	8	12	15	18	10	3

17. Copy and complete the following table to correspond with the histogram shown in Fig. 4.

Number of marbles	0 –	5 –	15 – 35
Frequency			200

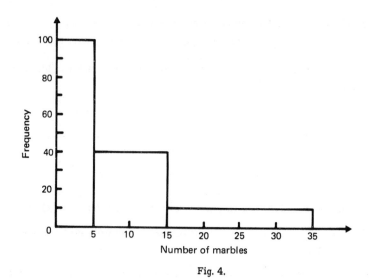

Fig. 4.

18. Write down the name of a distribution in which the arithmetic mean, the median and the mode all have the same value.

19. Calculate the mean deviation from the arithmetic mean of

5, 8, 8, 8, 8, 11

20.

Length (cm)	30 –	32 –	34 –	etc.
Frequency	4	8	14	etc.

Write down the limits of the second class if the lengths were measured

(a) to the nearest centimetre,　　(b) to the nearest millimetre.

21. The standard deviation of the numbers $x_1, x_2, \ldots, x_n$ is 8. Determine the standard deviation of

(a) $2x_1, 2x_2, \ldots, 2x_n$

(b) $3x_1 + 1, 3x_2 + 1, \ldots, 3x_n + 1$

22. The table below shows the wages of employees of a small firm. Estimate the percentage of employees earning between £42 and £66 per week.

Weekly wage (£)	36 –	48 –	60 –	75 –	90 – 150
Frequency	11	25	24	13	4

23. On a beach, 100 areas of equal size were selected and the numbers of a particular shellfish counted with the results shown below. Determine the median.

No. of shellfish	0	1	2	3	4	5	6	7
No. of areas	30	25	20	15	4	3	2	1

24. The number of flowers on a particular species of shrub is shown in the table below. Write down the mode of this distribution.

No. of flowers	6	7	8	9	10	11	12	13	14	15
No of shrubs	2	3	3	5	9	12	8	6	3	1

25. 200 packets of soap powder were weighed. It was found that the weights differed slightly. The lightest packet is replaced by an even lighter packet. Name a measure of dispersion which would be exactly the same before and after the replacement.

26. Write down the median of the following distribution.

No. of children	0	1	2	3	4
No. of families	38	30	22	8	2

27. Calculate the standard deviation of the following distribution.

x	8	12	16	20
Frequency	120	170	100	10

28. A group of people have the following age distribution:

Age (years)	20 – 23	24 – 27	28 – 31	32 – 35	36 – 39
Frequency	4	28	20	10	3

Estimate the mean age giving your answer to the nearest whole number of years.

29. The following numbers are in ascending order:

$$-8, -4, -2, 0, x, y, z$$

If their mean is zero

(a) find the sum of x, y and z

(b) calculate the mean deviation from the mean of the seven numbers.

30. The arithmetic mean of the three numbers a, b and c is 10 whilst the mean of the five numbers a, b, c, x and y is 16. Calculate y if x is 2.

31. Draw a cumulative frequency curve (an ogive) corresponding to the frequency curve shown in Fig. 5.

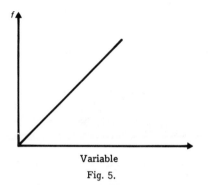

Variable

Fig. 5.

32. Fig. 6 shows the frequency curve of a moderately skewed distribution. The vertical drawn at A divides the area under the curve into two equal parts. Write down the name for the value of the variable at A.

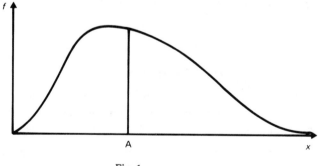

A

Fig. 6.

33. Estimate the mode of the following distribution.

Length (cm)	150 –	160 –	170 –	180 –	190 – 200
Frequency	100	128	88	50	34

34. Calculate the mean age of 100 children if the distribution of ages is as follows:

Age (years)	11	12	13	14
No. of children	20	40	30	10

35. Sketch a histogram to represent the frequency distribution corresponding to the frequency distribution curve shown in Fig. 7.

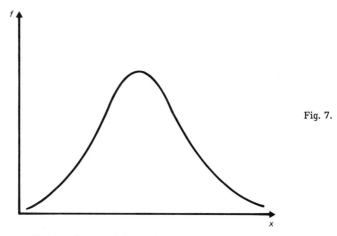

Fig. 7.

36. Sketch frequency distributions which are (a) bi-modal, (b) normal.

STANDARD QUESTIONS

37. 100 students took a test in mathematics. The table below shows the distribution of scores:

Score	Frequency
36 – 40	1
41 – 45	2
46 – 50	5
51 – 55	10
56 – 60	16
61 – 65	28
66 – 70	19
71 – 75	11
76 – 80	5
81 – 85	2
86 – 90	1

Construct a cumulative frequency curve and from it find

(a) the median score,

(b) the semi-interquartile range,

(c) the 70th percentile,

(d) the percentile corresponding to a score of 80,

(e) the number of pupils scoring more than 55 marks.

38. The examination marks obtained by 200 candidates were distributed as follows:

Mark	0 – 19	20 – 29	30 – 39	40 – 49
Frequency	16	14	28	46

Mark	50 – 59	60 – 69	70 – 79	80 – 89
Frequency	52	24	12	8

By using an assumed mean calculate the arithmetic mean and the standard deviation.

A further group of 400 candidates obtained a mean mark of 55.1 with a standard deviation of 14.8 marks. Determine the mean and standard deviation for the entire group of 600 candidates.

39. One hundred metal parts were produced on a certain machine. Their diameters were measured accurately and gave the following frequency distribution:

Diameter (mm)	260 –	265 –	270 –	275 –	280 –	285 –	290 – 295
No. of parts	1	5	25	45	20	3	1

(a) Construct an ogive (cumulative frequency curve) for this distribution. Use a scale of 2 cm to represent 10 parts and also to represent 5 mm of diameter.

(b) Using the ogive find the median and the semi-interquartile range of this distribution.

(c) A part is considered to be satisfactory if its diameter lies between 273 mm and 283 mm. Determine the number of parts which are not satisfactory.

40. An examination in German consisted of an oral part and a written part. On analysing the results of the examination the following results were obtained:

	Arithmetic mean	Standard deviation
Oral	43	17
Written	52	30

Each set of marks was now adjusted to an arbitrary scale with an arithmetic mean of 50 and a standard deviation of 20. The final mark was obtained by adding the two adjusted marks together. Calculate:

(a) the final mark for a candidate who scored 54 marks in the oral part and 41 in the written part

(b) the final mark for a candidate who scored 35 marks in the oral part and 73 in the written part.

41. A rectangular enclosure must be 30 metres by 25 metres, both measurements being correct to the nearest metre. Calculate:

(a) the minimum length of fencing required around the enclosure

(b) the maximum length of fencing required.

42. The table shows the age distribution of the population of a certain country in a certain year.

(a) Copy and complete the cumulative frequency table below. (Assume a maximum age 100 years.)

Age Group	Population (millions)	Cumulative Frequency
Under 10	18	
10 – 20	16	
20 – 30	14	
30 – 40	14	
40 – 50	14	
50 – 60	14	
60 – 70	10	
70 – 80	6	
80 and over	2	

(b) Draw the cumulative frequency graph using graph paper. (The group '10 – 20' means '10 but under 20 years', etc.)

(c) From your graph determine

(i) the median

(ii) the upper quartile

(iii) the lower quartile

(iv) the semi-interquartile range.

(d) Use your graph to estimate the number of people who were aged between 25 and 45 years in that year.

43. The table below shows the age distribution of the female population of Northern Ireland at 30 June 1978. Estimate the arithmetic mean and standard deviation for this distribution.

Age (years)	0 – 14	15 – 29	30 – 44	45 – 59	60 – 74	75 – 89
Number (thousands)	206	175	133	123	102	40

44. A biscuit manufacturer produces packets of cream crackers. A random sample of packets was weighed and the following figures recorded.

Net weight (grammes)	Number of packets
205 and less than 210	10
210 and less than 215	11
215 and less than 220	18
220 and less than 225	124
225 and less than 230	131
230 and less than 235	120
235 and less than 240	14
240 and less than 245	12

Determine the quartiles and median of this distribution. Hence find the semi-interquartile range.

45. Age of borrower purchasing a house by means of a mortgage in 1979.

Age of borrower	Number of heads of households
20 but under 25	21
25 but under 35	45
35 but under 45	21
45 but under 55	10
55 but under 65	3

For the above data determine the median, mean and standard deviation. Hence find a measure of skewness.

46. Purchase price of dwellings.

Price	Percentages	
	1970	1980
Under £9000	13	6
£9000 but under £11000	25	9
£11000 but under £12000	24	16
£12000 but under £15000	24	27
£15000 but under £20000	8	17
£20000 but under £30000	6	25

For the above data determine the median, mean and standard deviation. Hence find a measure of skewness.

47. The number of phone calls received in an office each hour over a period of 60 hours was recorded with the following results:

7	7	4	11	8	6	8	7	8	10
5	2	11	6	7	3	5	15	6	7
6	13	14	6	12	5	4	11	9	8
5	7	6	5	9	5	5	6	3	6
10	10	9	6	2	6	4	5	10	6
3	6	12	8	3	11	9	2	7	3

(a) Tabulate the information in a frequency distribution showing the frequency of each number of calls.

(b) What is the mode of the number of calls?

(c) Using the frequency distribution, calculate the mean number of calls received per hour.

48. A sample of eighty rods taken from the output of a production line was measured to the nearest mm and gave the following result:

Length (mm)	90 – 91	92 – 93	94 – 95	96 – 97	98
Number of rods	0	2	7	18	18

Length (mm)	99	100 – 101	102 – 105	106 – 109
Number of rods	19	12	4	0

(a) State the greatest and least possible lengths of a rod in the sample.

(b) Construct a cumulative frequency table and hence, taking 1 cm to represent 5 rods and also to represent 1 mm of length, draw a cumulative frequency curve for these data.

(c) Use your graph to estimate the median and the semi-interquartile range of the lengths of the rods.

(d) A rod is rejected if its length is less than 94.5 mm. Assuming that the sample is typical of the output, estimate the percentage of rods which have to be rejected.

49. Details of the annual bonuses paid to sales staff employed by a large wallpaper manufacturer have recently been made available (1979):

Annual bonus £	Number of bonuses
Under 60	8
60 and under 70	10
70 and under 80	16
80 and under 90	14
90 and under 100	10
100 and under 110	5

(Source: Company Records)

(a) Using graphical methods, estimate the median bonus paid and comment on the suitability of the median in interpreting this type of data. How many sales staff receive an annual bonus of

    (i) less than £88?   (iii) £96 or more?

    (ii) at least £63 but less than £75?

(b) From the graph, estimate a measure of dispersion, using the quartiles. What is meant by *dispersion?*

50. A daily count of the number of rejects from the assembly line of a local manufacturer has yielded the following data:

138	164	150	132	144	125	149	157
146	158	140	147	136	148	152	144
168	126	138	176	163	119	154	165
146	173	142	147	135	153	140	135
161	145	135	142	150	156	145	128

(a) Using the data, construct a frequency distribution table and from that sketch the corresponding frequency curve.

(b) Comment on the shape of the frequency curve you have obtained and compare it with the sketched shapes of *two* others with which you are familiar.

                     A First Course in Statistics

MULTIPLE CHOICE QUESTIONS

51. A histogram is

    a a measure of mass
    b a type of horizontal bar chart
    c a history of weights and measures
    d a diagram representing a frequency distribution.

52. The range of a distribution is

    a the smallest observation
    b the largest observation
    c the difference between the smallest and largest observations
    d another name for the mean deviation.

53. For the numbers 13, 18, 12, 11, 13, 19, 11, 16 and 11 the number 13 is

    a the mean   b the median   c the mode   d the range

54. The mean of four numbers is 14. Three of the numbers are 4, 10 and 16.
    What is the fourth number?

    a 10   b 36   c 30   d 26

55. The median of the numbers 4, 12, 6, 6, 8, 14 and 20 is

    a 6   b 8   c 10   d 70

56. Some steel bars are measured to the nearest millimetre. What is the upper limit
    of the class 130 mm but less than 150 mm?

    a 140 mm   b 149 mm   c 149.5 mm   d 150 mm

57. The diameters of some ball bearings are measured to the nearest 0.01 mm. The
    measurements are grouped into classes. What is the width of the class interval
    18.02 – 18.04?

    a 0.02 mm   b 0.03 mm   c 0.04 mm   d 0.01 mm

58. The information below relates to the mass in grammes of packets of a chemical.

Mass	Frequency
20	1
21	2
22	4
23	5
24	4
25	1

The mean of this distribution is

a 23   b 22.3   c 22.7   d 0.7

59. In question 58 the mode is

a 22   b 23   c 24   d $22\frac{2}{3}$

60. In question 58 the standard deviation is

a 22.7   b 5   c 23   d 1.3

61. The weekly wages of 30 employees of a firm were analysed with the following results:

Wage £	Frequency
48 – 56	5
56 – 64	12
64 – 72	10
72 – 80	3

The mean weekly wage is

a £58.96   b £2.96   c £62.93   d £60.74

62. In question 61 the modal class is

a 48 – 56   b 56 – 64   c 52   d 60

63. In question 61 the standard deviation is

a £1.74   b £7.00   c £32   d £34

64. Which of the following is a measure of central tendency?

a range

b standard deviation

c mode

d frequency

65. Which of the following is a measure of dispersion?

a mode

b range

c median

d coefficient of variation

66. The number 158 861 correct to 2 significant figures is

a 15   b 150 00   c 16   d 160 000

67. The number of 0.075 538 correct to 2 decimal places is

    a 0.076   b 0.075   c 0.07   d 0.08

68. The numbers 350 and 460 have been rounded to the nearest 10. The greatest possible value of 350 + 460 is

    a 810   b 820   c 800   d 811

69. The figures 290 and 240 have been rounded to the nearest 10. The least value of 290 − 240 is

    a 60   b 50   c 40   d 48

70. An error of 1 cm in measuring a length of 5 m gives a percentage error of

    a $\dfrac{1}{5}$ b $\dfrac{1}{500}$ c 20   d 0.2

71. The figures 22 and 35 are correct to 2 significant figures. The greatest possible value of 22 × 35 is

    a 798.75   b 741.75   c 770   d 828

72. One of the following is a discrete variable. Which?

    a the temperature of a room

    b the number of people at a football match

    c the quantity of liquid in a flask

    d the size of a machined component

73. The geometric mean of 9, 16 and 25 is

    a 16   b 60   c 16.7   d 15.3

74. The upper quartile of the numbers 2, 5, 3, 7, 4, 4, 6 is

    a 4   b 6   c 3   d 4.5

75. For the figures in question 74, the quartile deviation is

    a 1.5   b 3   c 5   d 2.5

76. Which of the frequency curves in Fig. 8 is likely to represent the distribution of the age of bridegrooms at marriage?

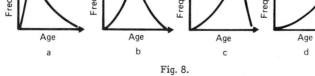

Fig. 8.

77. The type of frequency curve in Fig. 9 is called

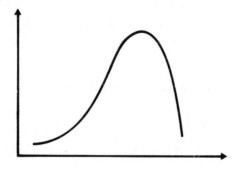

Fig. 9.

a skewed to the right

b skewed to the left

c normal

d bi-modal

78. The variance of the numbers 3, 4, 5 and 6 is

a 4.5   b 1.0   c 1.12   d 1.25

       A First Course in Statistics

7 STRAIGHT LINE GRAPHS

7.1 GRAPHS OF SIMPLE EQUATIONS

Consider the equation $y = 3x + 2$

As we have seen earlier on page 20 we can give x any value we please and hence calculate the corresponding value of y. Thus,

when $x = 0$ $y = 3 \times 0 + 2 = 2$

when $x = 2$ $y = 3 \times 2 + 2 = 8$

when $x = 4$ $y = 3 \times 4 + 2 = 14$

y depends upon the value allocated to x and is called the *dependent variable*. Since we can give x any value we please, x is called the *independent variable*. The values of the independent variable are plotted along the horizontal axis and this axis is frequently called the x-axis. The values of the dependent variable are then marked along the vertical axis which is often called the y-axis.

EXAMPLE 1. Draw the graph of $y = 2x + 3$ for values of x between 0 and 5.

Having decided on some values for x we calculate the corresponding values of y as follows:

when $x = 2$ $y = 2 \times 2 + 3 = 7$

when $x = 4$ $y = 2 \times 4 + 3 = 11$ and so on.

For convenience we tabulate corresponding values of x and y:

x	0	1	2	3	4	5
y	3	5	7	9	11	13

The graph is shown in Fig. 7.1 and it is seen to be a straight line. Equations of the type $y = 2x + 3$ where the highest power of the variables x and y is the first are called *simple equations*. All equations of this type give graphs which are straight lines and hence they are often called *linear equations*. In order to draw the graph of a linear equation we need only take two points. It is safer, however, to plot three points, the third point then acts as a check on the other two.

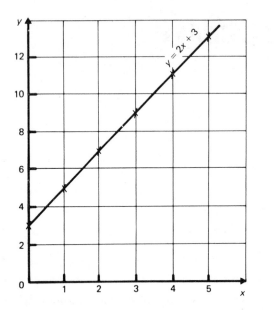

Fig. 7.1.

EXAMPLE 2. By means of a graph show the relationship between x and y in the equation $y = 5x + 2$. Plot the graph between $x = 2$ and $x = 7$

Since this is a linear equation we need only plot three points.

x	2	4	7
y	12	22	37

The graph is shown in Fig. 7.2.

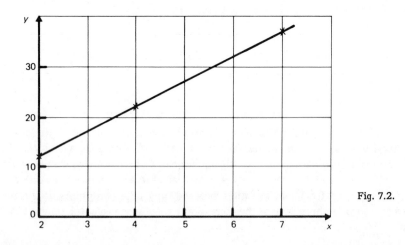

Fig. 7.2.

7.2 EQUATION OF A STRAIGHT LINE

Every linear equation may be written in the standard form

$$y = a + bx$$

Hence $y = 3 + 4x$ is in the standard form with $a = 3$ and $b = 4$.

The equation $y = 4 - 3x$ is in the standard form with $a = 4$ and $b = -3$.

7.2.1 Meaning of a and b in the Equation of a Straight Line

The point B is any point on the straight line shown in Fig. 7.3 and it has the coordinates x and y. Point A is where the straight line cuts the y-axis and it has the coordinates $x = 0$ and $y = a$.

$\dfrac{BC}{AC}$ is called the gradient of the line

Now, $BC = \dfrac{BC}{AC} \times AC = AC \times$ the gradient of the line

$y = BC + CD = BC + AO = AC \times$ gradient of the line $+ AO$

$y = x \times$ gradient of the line $+ a$

But $y = bx + c$ and hence it can be seen that

a is the intercept of the y-axis   **b is the gradient of the line.**

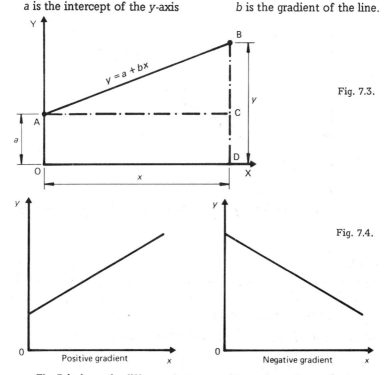

Fig. 7.3.

Fig. 7.4.

Fig. 7.4. shows the difference between positive and negative gradients.

7.3   FINDING THE EQUATION OF A STRAIGHT LINE

We are often faced with the problem of finding an equation connecting two variables whose graph is a straight line. The way in which we do this is shown in the next example.

EXAMPLE 3. The following values of x and y are thought to follow an equation of the type $y = a + bx$. Show that this is so and hence find values for a and b.

x	2	3	5	8
y	8	11	17	26

The graph is shown in Fig. 7.5 and it is a straight line. Hence the equation connecting x and y must be of the type $y = a + bx$.

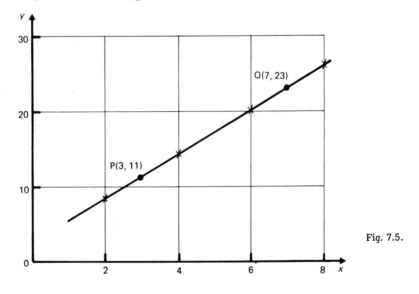

Fig. 7.5.

Method 1

To find the values of a and b, choose two points which lie on the line and find their coordinates. In Fig. 7.5, the point Q has the coordinates $x = 7$ and $y = 23$ whilst the point P has the coordinates $x = 3$ and $y = 11$.

Substituting the values of these coordinates in the equation $y = a + bx$

for point Q we have $\quad 23 = a + 7b \quad\quad (1)$

for point P we have $\quad 11 = a + 3b \quad\quad (2)$

Subtracting equation (2) from equation (1) gives

$$12 = 4b$$

$$b = \frac{12}{4} = 3$$

Substituting $b = 3$ in equation (1) gives

$$23 = a + 7 \times 3$$

$$23 = a + 21$$

$$a = 23 - 21$$

$$a = 2$$

Hence the equation connecting x and y is

$$y = 2 + 3x$$

Method 2

The value of b may be found by drawing a right-angled triangle ABC as shown in Fig. 7.6. Thus

$$b = \frac{BC}{AC} = \frac{12}{4} = 3$$

Then by taking any point that lies on the straight line and finding its coordinates the value of a may be found. In Fig. 7.6, A(2,8) has been chosen. We have

$$y = a + bx$$

$$8 = a + 3 \times 2$$

$$8 = a + 6$$

$$a = 2$$

Hence the equation of the straight line is $y = 2 + 3x$.

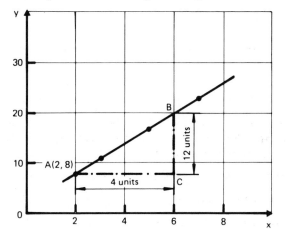

Fig. 7.6.

7.4 SCATTER DIAGRAMS

One of the most important applications of the straight line equation is the determination of an equation connecting two variables when their values have been obtained from a statistical enquiry. Usually when the graph is plotted it will be found that the points deviate slightly from a straight line. Since the data were obtained from a statistical enquiry we must expect this to happen. The graph depicting the result of the enquiry is called a *scatter diagram.*

It can be shown that the best straight line passes through the point $(\bar{x}, \bar{y})$ where

$\bar{x}$ is the arithmetic mean of the x values

$\bar{y}$ is the arithmetic mean of the y values

EXAMPLE 4. The following table gives the mean daily temperature, T (in °C) and the amount of electricity consumed, C (in MW) on seven consecutive Mondays during a certain year.

T	3	0	2	4	7	6	9
C	3.7	3.8	3.7	3.6	3.4	. 3.5	3.3

Plot a graph with T taken on the horizontal axis and confirm that the data approximates to a straight line. Hence find the equation connecting T and C.

The first step is to plot the points on a scatter diagram (Fig. 7.7). It will be

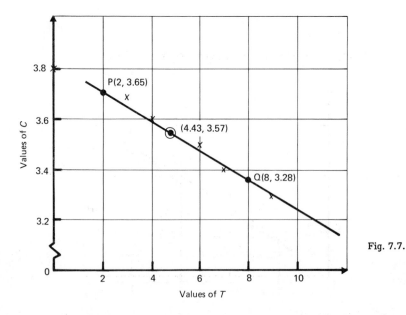

Fig. 7.7.

noticed that the points deviate only slightly from a straight line. To draw the straight line approximating to the plotted points we first calculate the values of $\bar{T}$ and $\bar{C}$.

$$\bar{T} = \frac{3 + 0 + 2 + 4 + 7 + 6 + 9}{7} = \frac{31}{7} = 4.43$$

$$\bar{C} = \frac{3.7 + 3.8 + 3.7 + 3.6 + 3.4 + 3.5 + 3.3}{7} = \frac{25}{7} = 3.57$$

This point is now plotted on the scatter diagram and the best straight line is now drawn. Although the straight line will not pass through some of the points an attempt must be made to ensure an even spread of points above and below the line as has been done in the diagram.

Since the variables give a straight line they must be connected by an equation of the type

$$C = a + bT$$

Note that since T has been taken on the horizontal axis it is the independent variable.

To obtain the values of the constants a and b, we use either of the methods shown in Example 3. Using the simultaneous equation method we choose two points which lie on the line. Choose the points a fair distance apart as this will help the accuracy of the result.

The point P (2, 3.65) lies on the line. Hence

$$3.65 = a + 2b \qquad (1)$$

The point Q (8, 3.28) lies on the line. Hence

$$3.28 = a + 8b \qquad (2)$$

Subtracting equation (2) from equation (1) gives

$$0.37 = -6b$$

$$b = -\frac{0.37}{6} = -0.062$$

Substituting $b = -0.062$ in equation (1) gives

$$3.65 = a + 2 \times (-0.062)$$

$$3.65 = a - 0.124$$

$$a = 3.65 + 0.124 = 3.774$$

Hence the equation which approximately connects the temperature and the amount of electricity consumed is

$$C = 3.774 - 0.062T$$

7.5 NON-LINEAR EQUATIONS WHICH CAN BE REDUCED TO THE LINEAR FORM

Many non-linear equations can be reduced to the linear form by making a suitable substitution. Some common forms of non-linear equations are:

$$y = a + bx^2$$

$$y = a + b\sqrt{x}$$

$$y = a + \frac{b}{x}$$

$$y = a + \frac{b}{x^2}$$

If y is plotted against x, each of the above equations will give a graph which is a smooth curve (hence the term non-linear).

Consider the equation $y = a + bx^2$. If we let $z = x^2$, the equation becomes

$$y = a + bz$$

Since $y = a + bz$ is a linear equation, if we plot y against z we shall get a straight line graph. In effect y has been plotted against x^2 and we have reduced the non-linear equation, $y = a + bx^2$, to a linear form. The intercept on the y-axis is a and the gradient is b.

In a similar way the equation $y = a + \dfrac{b}{x}$ can be reduced to the linear form by plotting y against $\dfrac{1}{x}$. The gradient is then b and the intercept on the y-axis is a.

EXAMPLE 5. A manufacturer's production costs, £C, vary with the number produced per annum, N, as shown in the table below.

N	1000	2000	3000	4000	5000
C	7300	9900	12 000	13 600	15 100

It is expected that the relationship between N and C is of the form $C = a + b\sqrt{N}$ where a and b are constants. Plot C against $\sqrt{N}$ and draw the line of best fit. Hence estimate the values of a and b and use the relationship to predict the value of C when $N = 3400$.

The first step is to prepare another table giving corresponding values of $\sqrt{N}$ and C as shown on the following page.

$z = \sqrt{N}$	31.6	44.7	54.8	63.2	70.7
C	7300	9900	12 000	13 600	15 100

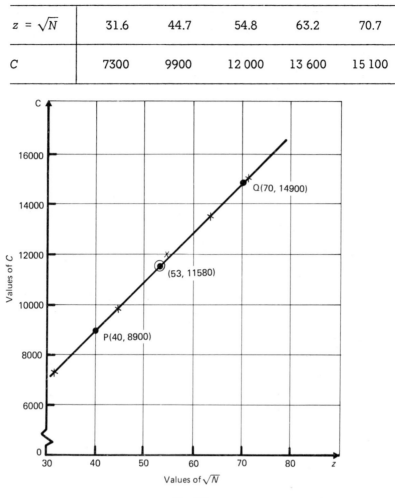

Fig. 7.8.

$\bar{z} = 53$ and $\bar{C} = 11\,580$

The graph is plotted in Fig. 7.8, and since it is a straight line $C = a + b\sqrt{N}$. To estimate the values of a and b choose two points P (40, 8900) and Q (70, 14 900) which lie on the line.

For the point Q: $14\,900 = a + 70b$ (1)

For the point P: $8\,900 = a + 40b$ (2)

Subtracting equation (2) from equation (1) gives

$6000 = 30b$

$b = \dfrac{6000}{30} = 200$

Substituting $b = 200$ in equation (1) gives

$$14\,900 = a + 70 \times 200$$

$$14\,900 = a + 14\,000$$

$$a = 14\,900 - 14\,000 = 900$$

Hence the equation connecting C and N is

$$C = 900 + 200\sqrt{N}$$

Using the relationship, we find that

when $N = 3400, C = 900 + 200\sqrt{3400}$

$$= 900 + 200 \times 58.3$$

$$= 900 + 11\,660 = 12\,560$$

Therefore when 3400 items are produced the cost of production is £12 560.

EXERCISE 13

1.   Draw the graphs of the following simple equations:

(a) $y = x + 2$ taking values of x between 0 and 6.

(b) $y = 3x + 7$ taking values of x between 2 and 7.

(c) $y = 9 - 3x$ taking values of x between 0 and 3.

2.   The following equations represent straight lines. State in each case the gradient of the line and the intercept on the y-axis:

(a) $y = x + 3$
(b) $y = 3x + 4$
(c) $y = 12 - 5x$

3.   Find the values of a and b if the straight line $y = a + bx$ has a gradient of 4 and passes through the point $(2, 15)$.

4.   Find the values of a and b if the straight line $y = a + bx$ passes through the point $(3, 7)$ and the intercept on the y-axis is 1.

5.   The straight line $y = a + bx$ passes through the points $(1, 1)$ and $(2, 4)$. Find the values of a and b.

6.   The following table gives values of x and y which are connected by an equation of the type $y = a + bx$. Plot the graph and from it find values for a and b.

x	2	4	6	8	10	12
y	10	16	22	28	34	40

7. The following values of P and Q were obtained as a result of a statistical enquiry. P and Q are supposed to be related by an equation of the type $P = a + bQ$. Find by plotting the graph the most probable values of a and b.

Q	2.5	3.5	4.4	5.8	7.5	9.6	12.0	15.1
P	13.6	17.6	22.2	28.0	35.5	47.4	56.1	74.6

8. A record is kept of maintenance cost for several identical automatic machines. The table below gives details.

Age T (years)	6	2	7	5	3	1
Cost C (£)	135	48	148	110	73	30

T and C are expected to be related by an equation of the type $C = a + bT$. Check, by plotting a graph of T against C, if this is so and estimate the values of a and b. Using the relationship obtained find the probable maintenance cost for a machine which is 4 years old.

9. It is claimed that the demand for goods transport appears to be related to the gross domestic product of a country. The data below show estimated tonne − kilometres of goods traffic in Great Britain at 1970 factor cost. The total tonne − kilometres, M, and the gross domestic product, P, are thought to be connected by an equation of the type $M = a + bP$. By drawing a graph of M against P show that this is so, and hence estimate values for a and b.

Year	1964	1965	1966	1967	1968	1969	1970
P (£1000m)	38	39	42	43	46	47	48
M (tonne-km)	47	48	50	51	53	54	55

10. The table below gives details of production output (P units) and total costs (£T). Draw a graph of these data and show that P and T are connected by an equation of the type $T = a + bP$ and find values for a and b.

Week number	1	2	3	4	5	6	7	8	9	10	11	12
P	47	53	49	58	44	57	59	60	65	52	48	56
T	98	110	100	122	90	115	124	125	133	110	98	114

11. The following values of x and y satisfy an equation of the type $y = a + bx^2$. Plot a graph of y against x^2 and hence find values for a and b.

x	2	3	4	5	6
y	22	45	76	115	162

12. The values tabulated below are thought to obey a law of the type $R = a + b\sqrt{Q}$. By plotting R (vertically) against $\sqrt{Q}$ (horizontally) show that this is so and hence find values for a and b.

Q	1	4	9	16	25
R	2	3.5	5	6.5	8

13. In a production process the cost of producing a unit £C depends upon the number produced N as shown in the table below. By plotting C against $\dfrac{1}{N}$ show that N and C are connected by an equation of the type $C = a + \dfrac{b}{N}$ and hence find values for a and b. Using the relationship estimate the cost of a unit when 30 are produced.

N	2	5	10	20	40
C	450	330	290	270	254

14. During a statistical enquiry the following values of two variables A and B were obtained.

A	1	2	3	4	5
B	98	25	12	6	4

Show that A and B are connected by an equation of the type $B = \dfrac{k}{A^2} + C$ by plotting a graph of B against A^2. Hence find values of k and C and use them to estimate the value of B when A is 2.6.

8 REGRESSION AND CORRELATION

8.1 CURVE FITTING

Readings which are obtained as a result of a statistical enquiry or experiment usually contain errors in observation and measurement. When the points are plotted on a graph it is usually possible to visualise a straight line or curve which approximates to the data. Thus in Fig. 8.1 the data appear to be approximated by a straight line whilst in Fig. 8.2 the data are approximated by a curve.

Figs. 8.1 and 8.2 are called *scatter diagrams*. The problem is to find equations of curves or straight lines which approximately fit the plotted data. Finding equations of approximating curves or straight lines is called *curve fitting*. In this book only approximating straight lines will be considered.

It was shown in Chapter 7 that the best straight line approximating to plotted data passes through the point $(\overline{x}, \overline{y})$, where x is the independent variable and y is the dependent variable and $\overline{x}$ and $\overline{y}$ are the arithmetic means of the x and y values respectively. This straight line has the equation

$$y = a + bx$$

where a and b are constants whose values may be determined by either of the methods shown in Chapter 7.

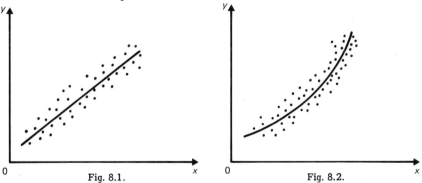

Fig. 8.1.   Fig. 8.2.

8.2 REGRESSION

Suppose we are given several corresponding values of x and y which, when plotted, approximate to a straight line. To find a value for y corresponding to a stated value of x (which is not included in the given data) we first obtain the equation of the line which best fits the given data. It will be of the type

$$y = a + bx$$

This line is called the regression line of y on x because y is estimated from x.

Sometimes we wish to obtain the value of x corresponding to a stated value of y.

In this case we use the regression line of x on y which will be of the type

$$x = a_1 + b_1 y$$

a_1 and b_1 being constants. Generally the regression line of y on x is not the same as the regression line of x on y. However, both these regression lines pass through the point $(\bar{x}, \bar{y})$.

8.3   CORRELATION

When corresponding values of two variables x and y obtained by experiment are plotted a scatter diagram like that shown in Fig. 8.1 is obtained. A rough relationship, or *correlation*, is seen to exist beween x and y.

Correlation is closely associated with regression. We seek to determine how well a linear (or other) equation describes the relationship between the two variables. When the points on a scatter diagram are such that they approximate to a straight line, the correlation is said to be linear. Only linear correlation will be considered.

The correlation may be positive, precise or negative. For positive correlation (Fig. 8.3) large values of y accompany large values of x. For the correlation to be precise (Fig. 8.4) all the points on the scatter diagram must lie on a straight line. For negative correlation (Fig. 8.5) the values of y decrease as the values of x increase. If no relationship is indicated by the points on the scatter diagram we say that there is no correlation between the variables x and y (Fig. 8.6).

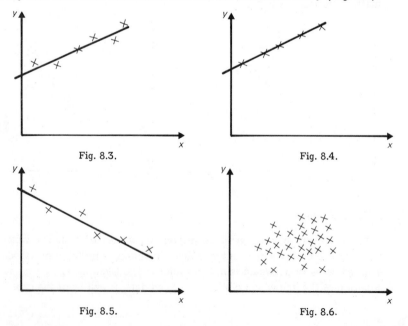

Fig. 8.3.   Fig. 8.4.

Fig. 8.5.   Fig. 8.6.

8.4 MEASURE OF CORRELATION

We have seen that the regression line for y on x is given by the equation

$$y = a + bx$$

and that the regression line for x on y is given by the equation

$$x = a_1 + b_1 y$$

These two equations are identical only if the correlation is precise, that is, only if all the points on the scatter diagram lie on a straight line.

We can often see, by direct observation of the scatter diagram, how well a straight line describes the relationship between two variables. In Fig. 8.3 we see that the straight line describes the relationship between x and y very well indeed. In Fig. 8.1, however, the relationship is not so well defined by the straight line.

Hence in order to deal with the problem of scatter we need a measure of correlation.

8.5 COEFFICIENT OF CORRELATION

When the equations of the regression lines of y on x and x on y are known the coefficient of regression may be calculated from

$$r = \sqrt{bb_1}$$

EXAMPLE 1. The regression line of y on x is $y = 4.15 + 0.153x$ and the regression line of x on y is $x = -27.03 + 6.154y$. Calculate the coefficient of correlation. We are given that $b = 0.153$ and that $b_1 = 6.154$. Hence

$$r = \sqrt{0.153 \times 6.154} = 0.9983$$

It can be shown that the value of the correlation coefficient lies between $+1$ and -1. When $r = +1$ precise positive correlation exists and when $r = -1$ precise negative correlation occurs. When $r = 0$ no correlation exists.

In practice, values of zero or unity occur very rarely and usually r lies between 0.6 and 0.9. The higher the value of r the better is the correlation. If the value of r lies between $+0.3$ and -0.3 the correlation is regarded as being unimportant, i.e. there is practically no correlation. However the degree of correlation depends upon the value of r and the number of observations considered. Fig. 8.7 shows the values of r needed for good and excellent correlation for 5, 10 and 20 observations.

EXAMPLE 2. The following table shows corresponding values of two variables x and y.

x	2	3	5	8	10
y	5	7	11	17	21

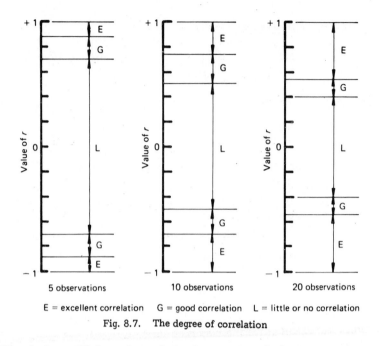

E = excellent correlation   G = good correlation   L = little or no correlation
Fig. 8.7.   The degree of correlation

(a) What is the coefficient of correlation between x and y?
(b) What is the value of the regression coefficient for the regression line of y on x?

(a) On drawing the graph (Fig. 8.8) we see that all the points lie on a straight line. Hence the correlation between x and y is precise and therefore the

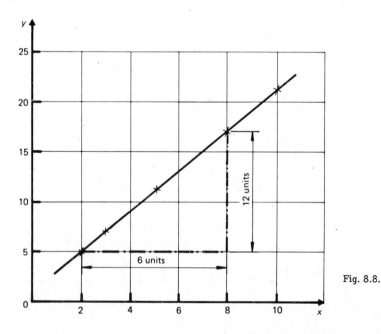

Fig. 8.8.

coefficient of correlation between x and y is $+1$, the plus sign indicating that the straight line slopes upwards to the right.

(b) The value of the regression coefficient for the regression line of y on x is the value of the gradient of the line. Thus

$$\text{gradient} = \frac{12}{6} = 2$$

So the value of the regression coefficient for the regression line of y on x is 2.

8.6 RANK CORRELATION

Instead of using precise values of the two variables the data may be ranked in order of size from the greatest down to the smallest, or vice versa. When two variables are ranked in this way the coefficient of rank correlation, known as Spearman's coefficient, is given by:

$$R = 1 - \frac{6\Sigma D^2}{N(N^2 - 1)}$$

where D = the difference between ranks for corresponding values of the two variables and N = the number of pairs of observations in the data.

EXAMPLE 3. Two judges were asked to rank six samples of ice cream, A, B, C, D, E and F, in order of their preference. They submitted the choices shown below:

	A	B	C	D	E	F
1st judge (x)	2	5	1	4	3	6
2nd judge (y)	5	4	3	2	1	6

Calculate the coefficient of rank correlation.

x	y	$D = x - y$	D^2
2	5	-3	9
5	4	1	1
1	3	-2	4
4	2	2	4
3	1	2	4
6	6	0	0
		$\Sigma D^2 = 22$	

$$R = 1 - \frac{6 \times 22}{6(6^2 - 1)}$$

$$= 1 - \frac{132}{210}$$

$$= 1 - 0.63$$

$$= 0.37$$

If the two sets of ranks are equal in every respect then $\Sigma D^2 = 0$ and $R = +1$. This indicates perfect correlation of the two sets of ranks. If there is lack of correlation $R = -1$, that is, if the 1st rank of one set corresponds to the last rank of the other set, the 2nd rank of the first set corresponds with the next to last rank of the second set, and so on. Intermediate values of R may be taken as having approximately the same meaning as the values of the correlation coefficient, r.

In the case of Example 3 the value of R is low and we can say that the two judges do not agree very well.

EXAMPLE 4. The information below relates to the performance of ten students in examinations. Find the coefficient of rank correlation and comment on the result.

Student number	1	2	3	4	5	6	7	8	9	10
Marks in maths	12	15	8	10	11	10	7	17	6	5
Mark in statistics	10	11	7	12	9	8	7	14	2	4

The first step is to arrange the data in order of rank.

Student number	1	2	3	4	5	6	7	8	9	10
Rank in maths (x)	3	2	7	5.5	4	5.5	8	1	9	10
Rank in statistics (y)	4	3	7.5	2	5	6	7.5	1	10	9

The calculation of the coefficient of rank correlation is then as follows:

x	y	$D = x - y$	D^2
3	4	-1	1
2	3	-1	1
7	7.5	-0.5	0.25
5.5	2	3.5	12.25
4	5	-1	1
5.5	6	-0.5	0.25
8	7.5	0.5	0.25
1	1	0	0
9	10	-1	1
10	9	1	1
		$\Sigma D^2 = 18.00$	

$$r = 1 - \frac{6 \times 18}{10(10^2 - 1)}$$

$$= 1 - \frac{108}{990}$$

$$= 1 - 0.109$$

$$= 0.891$$

Note the way in which we deal with the equal ranks. Considering the marks in mathematics, the ranks are:

Rank	Mark
1	17
2	15
3	12
4	11
Equal { 5	10
5th { 6	10
7	8
8	7
9	6
10	5

10 marks is ranked equal 5th and students 4 and 6 therefore ranked 5.5 (i.e. the arithmetic mean of 5 and 6). Similarly when dealing with the marks in Statistics, students 3 and 7 are ranked joint 7th and for purposes of calculating the rank correlation coefficient they are each ranked 7.5, the arithmetic mean of 7 and 8.

Using Fig. 8.7, for 10 observations, we see that the correlation is excellent. There is excellent correlation between the marks in mathematics and statistics.

EXAMPLE 5. The weights, correct to the nearest kilogram, and the heights, correct to the nearest centimetre, of 12 adults are given in the following table.

Adult	A	B	C	D	E	F	G	H	I	J	K	L
Weight (x)	74	70	55	58	51	56	57	62	62	69	68	67
Height (y)	175	173	164	171	152	164	155	169	168	178	186	175

(a) Calculate Spearman's coefficient of rank correlation between height and weight.

(b) Plot a scatter diagram of the data.

(c) Calculate the arithmetic mean values of weight and height and use them to draw the line of best fit.

(d) Obtain the equation of this line in the form $y = a + bx$ where a and b are constants. Use it to estimate the weight when the height is 160 cm.

(a)

Adult	Rank in weight (R_1)	Rank in height (R_2)	$D = R_1 - R_2$	D^2
A	12	9.5	2.5	6.25
B	11	8	3.0	9.00
C	2	3.5	− 1.5	2.25
D	5	7	− 2.0	4.00
E	1	1	0	0.00
F	3	3.5	− 0.5	0.25
G	4	2	2.0	4.00
H	6.5	6	0.5	0.25
I	6.5	5	1.5	2.25
J	10	11	− 1.0	1.00
K	9	12	− 3.0	9.00
L	8	9.5	− 1.5	2.25

$$R = 1 - \frac{6\Sigma D^2}{N(N^2 - 1)} = 1 - \frac{6 \times 40.5}{12 \times 143} = 0.8584 \qquad 40.50$$

(b) The scatter diagram is shown in Fig. 8.9.

(c) $\bar{x} = \dfrac{74 + 70 + 55 + 58 + 51 + 56 + 57 + 62 + 62 + 69 + 68 + 67}{12}$

$= \dfrac{749}{12} = 62.4$

$\bar{y} = \dfrac{175 + 173 + 164 + 171 + 152 + 164 + 155 + 169 + 168 + 178 + 186 + 175}{12}$

$= \dfrac{2030}{12} = 169.2$

(d) To find the equation of the line of best fit the right-angled triangle ABC (Fig. 8.9) has been drawn. By scaling, AC = 15 and BC = 16. Hence

$$\text{gradient} = b = \frac{16}{15} = 1.07$$

Since the line of best fit is a straight line its equation is of the type

$y = a + bx$

To find the value of a we use the coordinates of the point A(55, 161):

$161 = a + 1.07 \times 55$

$161 = a + 59$

$a = 102$

$\therefore y = 102 + 1.07x$

We now have to find x when $y = 160$.

$$160 = 102 + 1.07x$$

$$1.07x = 160 - 102$$

$$x = \frac{58}{1.07} = 54$$

Hence when the height is 160 cm the estimate of the weight is 54 kg.

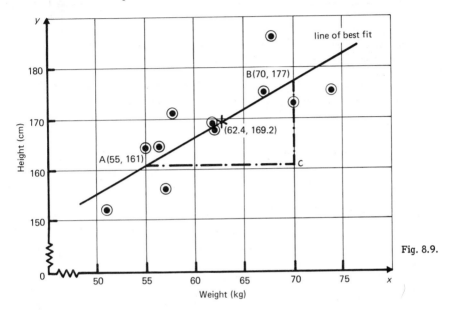

Fig. 8.9.

EXERCISE 14

1. The regression line of y on x is $y = 5 + 4x$ and the regression line of x on y is $x = 0.21y - 1.5$. Calculate the value of the correlation coefficient. If there were ten observations, use the diagrams of Fig. 8.7 to decide if the correlation between x and y is excellent or good or if there is little or no correlation.

2. Which of the diagrams shown in Fig. 8.10 indicates negative correlation?

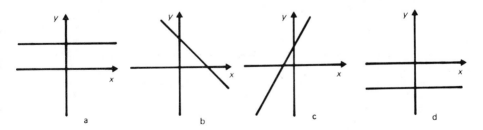

Fig. 8.10.

3. Choose from the numbers 1.2, 0.8, 0.4, 0.002, − 0.6, − 0.95, − 1.3, the most suitable correlation coefficient to match the diagrams a, b and c in Fig. 8.11.

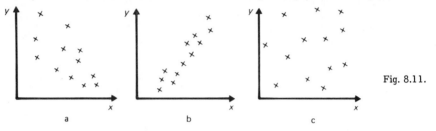

Fig. 8.11.

a   b   c

4. Which one of the following correlation coefficients shows the least correlation: −0.8, −0.4, 0, 0.3?

5. Copy the diagram (Fig. 8.12) and insert points which will indicate that a large negative linear correlation exists between the two variables x and y.

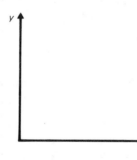

Fig. 8.12.

6. The table below shows corresponding values of two variables x and y.

x	3	4	6	9	10
y	5	7	11	17	19

Write down the value of the correlation coefficient between x and y.

7. The following table shows corresponding values of two variables P and Q, P being the independent variable.

P	10	12	15	20	30
Q	80	76	70	60	40

What is the value of the correlation coefficient between P and Q?

8. A set of observations of a pair of variables x and y is such that $\bar{x} = 10$ and $\bar{y} = 6$. If the line of regression of y on x passes through the point x = 8,

$y = 20$ determine: (a) the coefficient of regression of y on x, (b) the intercept that the line of regression of y on x makes on the y-axis.

9. From a group of observations of a pair of variables, x and y, it is found that the regression coefficient of y on x is -0.5. The line of regression passes through the point $(1, 8)$. Estimate the value of y when $x = 7$.

10. The marks of ten pupils studying languages were:

German	50	24	60	68	66	39	44	34	49	76
French	59	38	65	71	68	50	53	44	52	77

(a) Draw a scatter diagram of the marks plotting German marks horizontally and French marks vertically.

(b) Calculate the mean German mark, call it $\bar{g}$.

(c) Calculate the mean French mark, call it $\bar{f}$.

(d) Plot the point having coordinates $(\bar{g}, \bar{f})$.

(e) Draw the straight line which best fits the data and passes through the point $(\bar{g}, \bar{f})$.

(f) Find the equation of the line of best fit in the form $f = a + bg$ where f indicates the French mark, g indicates the German mark, and a and b are constants.

11. The following items were ranked in ascending order of magnitude so that a rank correlation coefficient may be calculated: 4, 8, 5, 6, 10, 10, 21, 15, 22. What would be the rank of the two items of magnitude 10?

12. The table below shows the marks obtained and the resulting rank for ten pupils in two mathematics tests. Copy and complete the last two columns of this table.

Pupil	Test 1	Test 2	1st test rank (R_1)	2nd test rank (R_2)	$D = R_1 - R_2$	D^2
A	50	65	9	1	8	64
B	43	55	10	3		
C	56	56	7	2	5	25
D	52	53	8	4		
E	68	52	5	5	0	0
F	66	49	6	6	0	0
G	74	48	3	7		
H	70	45	4	8	-4	16
I	80	42	1	9		
J	76	40	2	10	-8	64

$$\Sigma D^2 = 314$$

(a) Illustrate the two sets of marks by means of a scatter diagram plotting the Test 1 marks on the x-axis.

(b) Draw the line of best fit and find its equation in the form $y = a + bx$, y representing the Test 2 marks and x the Test 1 marks.

(c) Using the formula $r = 1 - \dfrac{6\Sigma D^2}{N(N^2 - 1)}$ find the rank correlation between the performance in Test 1 and the performance in Test 2.

13. The table below shows the marks awarded to six divers by two judges. Calculate the value of a coefficient of rank correlation.

Diver	A	B	C	D	E	F
1st judge	6.5	7.0	7.0	8.0	8.5	9.0
2nd judge	7.5	9.0	8.0	6.5	8.5	8.5

14. The table below shows the marks awarded to six competitors in a poetry competition by two judges. Calculate Spearman's coefficient of rank correlation.

Competitor	A	B	C	D	E	F
Judge X	19	16	15	14	12	10
Judge Y	14	18	7	8	17	10

15. Test for relative darkness of ten shades of red:

Correct rank order	1	2	3	4	5	6	7	8	9	10
Applicant 'A' rank order	2	4	1	3	5	7	6	9	8	10

(a) Find the rank correlation coefficient of these figures.

(b) Applicant 'B' also takes the test.

Applicant 'B' rank order: 5 1 2 3 4 10 6 7 8 9

Find the rank correlation coefficient between Applicant B's rank order and the correct order.

(c) On the basis of these results only, which applicant is more able to distinguish between shades of red?

16. Twelve samples of home made toffee were judged by two judges A and B who awarded marks out of a maximum of 100. The results are shown in the table.

Sample	1	2	3	4	5	6	7	8	9	10	11	12
Judge A	51	39	43	28	30	68	59	64	25	47	32	44
Judge B	60	45	50	35	35	80	70	75	30	55	40	55

(a) Draw a scatter diagram of the marks awarded using a scale of 2 cm to 10 marks. Let Judge A be represented by the horizontal scale and Judge B by the vertical scale.

(b) Calculate the mean marks awarded by each Judge and plot the results on the scatter diagram.

(c) Draw the line of best fit.

(d) Two further samples were judged by one judge only. Judge B gave the 13th samples 65 marks, while Judge A gave the 14th sample 71. Use your graph to estimate the marks that would most likely have been awarded by the other judge.

9 STANDARDISED RATES

9.1 CRUDE DEATH RATE

The crude death rate for a town, city or district is calculated from

$$\frac{\text{number of deaths}}{\text{population}} \times 1000$$

EXAMPLE 1. A town has a population of 20 000 people. In 1975 the total number of deaths was 300. Calculate the crude death rate.

$$\text{Crude death rate} = \frac{300}{20\,000} \times 1000 = 15$$

Hence the crude death rate is 15 deaths per 1000 of population.

9.2 STANDARDISED DEATH RATE

Between one town and another or one district and another the crude death rate provides no real basis for comparison. This is because it does not take into account the different age compositions of the two places. The crude death rates are therefore standardised by making use of a standard population as shown in the next example.

EXAMPLE 2. Find the crude and standardised death rates for the data given below.

Age range (years)	Population	Number of deaths	Standard population
0 –	3000	24	320
20 –	4000	12	260
40 –	4000	52	240
60 and over	2000	160	180

Note that the standard population gives the number of people in each age group per 1000 of standard population. The calculation for the standardised death rate is then as shown in the table on the following page.

Age range (years)	Specific death rate	Standard population	Deaths per 1000 of standard population
0 –	$\dfrac{24}{3000} \times 1000 = 8$	320	$\dfrac{8}{1000} \times 320 = 2.56$
20 –	$\dfrac{12}{4000} \times 1000 = 3$	260	$\dfrac{3}{1000} \times 260 = 0.78$
40 –	$\dfrac{52}{4000} \times 1000 = 13$	240	$\dfrac{13}{1000} \times 240 = 3.12$
60 and over	$\dfrac{160}{2000} \times 1000 = 80$	180	$\dfrac{80}{1000} \times 180 = 14.40$
		1000	20.86

Total population $= 3000 + 4000 + 4000 + 2000 = 13\,000$

Total number of deaths $= 24 + 12 + 52 + 160 = 248$

Crude death rate $= \dfrac{248}{13\,000} \times 1000 = 19.08$

Hence the crude death rate is 19.08 deaths per 1000 of population

Standardised death rate $= 2.56 + 0.78 + 3.12 + 14.40 = 20.86$

Hence the standardised death rate is 20.86 per 1000 of population.

In practice, male and female death rates are computed separately because death rates vary considerably as between males and females for the different age groups.

The standard population is determined from the population of the country as a whole as shown in Example 3.

EXAMPLE 3. The table below shows the number of deaths occurring in a particular year in each of two towns A and B, together with the population of each town and of the country as a whole; all are classified by age group.

Age group (years)	Number of deaths		Population		
	Town A	Town B	Town A	Town B	Country (millions)
0 –	42	96	6000	16 000	15
20 –	14	30	7000	15 000	14
40 –	64	154	5000	7000	13
60 and over	176	160	2000	2000	8

Calculate the crude and standardised rates for each town.

The first step is to express the population of the country as a standard population per 1000 people.

Total population of country = 15 + 14 + 13 + 8 = 50 millions.

Age group	Standard population
0 –	$\dfrac{15}{50}$ × 1000 = 300
20 –	$\dfrac{14}{50}$ × 1000 = 280
40 –	$\dfrac{13}{50}$ × 1000 = 260
60 and over	$\dfrac{8}{50}$ × 1000 = 160

The calculation of the crude and standardised death rates is then as follows:

Age range (years)	Town A Specific rate	Town A Deaths per 1000	Town B Specific rate	Town B Deaths per 1000	Standard population
0 –	7.0	2.10	6.0	1.80	300
20 –	2.0	0.56	2.0	0.56	280
40 –	12.8	3.33	22.0	5.72	260
60 and over	88.0	14.08	80.0	12.80	160
		20.07		20.88	1000

Town A: Standardised death rate = 20.07 deaths per 1000 of population

$$\text{Crude death rate} = \frac{296}{20\,000} \times 1000 = 14.8 \text{ deaths per 1000 of population}$$

Town B: Standardised death rate = 20.88 per 1000 of population.

$$\text{Crude death rate} = \frac{440}{40\,000} \times 1000 = 11 \text{ deaths per 1000 of population.}$$

9.3 STANDARDISED RATES APPLIED TO OTHER PROBLEMS

Standardised rates may be applied to problems other than death rates. They may be applied to any situation in which the population is classified into age groups. Some examples are: fertility rates, unemployment rates and accident rates.

EXAMPLE 4. The data below relate to the number of accidents in two factories. Calculate the crude and standardised accident rates.

Age group (years)	Factory A		Factory B	
	Number of employees	Number of accidents	Number of employees	Number of accidents
Under 20	300	25	200	18
20 –	1500	108	800	60
40 –	1400	109	750	61
60 and over	300	25	50	1

To determine the standardised accident rate in each factory we have to define a standard population. One way would be to use the age distribution for the industry in question. A second way, which will be used in this problem, is to use the joint population of both factories.

Age group (years)	Total number of employees in both factories	Standard population
Under 20	500	$\dfrac{500}{5300} \times 1000 = 94$
20 –	2300	$\dfrac{2300}{5300} \times 1000 = 434$
40 –	2150	$\dfrac{2150}{5300} \times 1000 = 406$
60 and over	350	$\dfrac{350}{5300} \times 1000 = 66$
	5300	1000

Having determined a standard population, we calculate the standardised rates as follows:

		Factory A		Factory B	
Age group (years)	Standard population	Specific rate	Accidents per 1000	Specific rate	Accidents per 1000
Under 20	94	$\frac{25}{300} \times 1000$ $= 83.33$	$\frac{83.33}{1000} \times 94$ $= 7.83$	$\frac{18}{200} \times 1000$ $= 90$	$\frac{90}{1000} \times 94$ $= 8.46$
20 –	434	72.00	31.25	75	32.55
40 –	406	77.86	31.61	81.33	33.02
60 and over	66	83.33	5.50	20	1.32
	1000		76.19		75.35

Crude accident rate in Factory A $= \dfrac{267}{3500} \times 1000 = 76.29$ per 1000 workers

Crude accident rate in Factory B $= \dfrac{140}{1800} \times 1000 = 77.78$ per 1000 workers

Although the crude rates show that factory B has a higher accident rate than factory A the standardised rates show the reverse, the figures being 76.19 for factory A and 75.35 accidents per 1000 workers for factory B. When comparing the accident rates for the two factories we use the standardised rates and we conclude, therefore, that factory B has a lower accident rate than factory A.

EXERCISE 15

1. A district has a population of 80 000 people. In a certain year the total number of deaths was 1280. Calculate the crude death rate.

2. In a town having a population of 20 000 the crude unemployment rate was 84 per 1000. Calculate the number actually unemployed.

3. In a certain year the number of children born to women in the age group 20 – 25 years was 29. If there were 367 women in that age group calculate the number of births per 1000 women in that age group (i.e. calculate the fertility rate).

4. Calculate the crude and standardised death rates for the following data.

Age group (years)	Population (thousands)	Number of deaths	Standard population
Under 10	22	360	225
10 –	29	101	300
25 –	35	230	291
45 –	16	356	148
65 and over	4	390	36

5. The figures below give information regarding the rate of unemployment in the country as a whole and in a certain district. Calculate:

(a) the standardised national rate of unemployment

(b) the crude unemployment rate for the district

(c) the standardised unemployment rate for the district.

Age group (years)	Under 30	30 –	45 –	60 and over
Standard population: Age constitution Unemployment rate (%)	250 6	350 7	300 11	100 15
Local population: Age constitution Unemployment rate (%)	300 5	350 8	300 12	50 20

6. The figures below show the number of accidents in two factories X and Y. For each factory calculate the crude and standardised accident rates.

Age group (years)	Under 25	25 –	45 –	65 and over
Factory X: Number of employees Number of accidents	350 20	600 52	250 29	300 7
Factory Y: Number of employees Number of accidents	600 32	800 104	400 37	200 8

10 INDEX NUMBERS

10.1 INTRODUCTION

An *index number* is a statistical measure designed to show changes over a period of time in the price, quantity or value of an item or a group of items.

10.2 PRICE RELATIVE

The simplest example of an index number is the *price relative* which is the ratio of the price of a commodity at a given time to its price at a different time. The ratio is usually expressed as a percentage.

EXAMPLE 1. In January 1970 the price of a certain commodity was 40 p whilst in January 1975 its price was 60 p. Taking January 1970 as the base period find the price relative.

$$\text{Price relative} = \frac{\text{price at January 1975}}{\text{price at base period}} \times 100$$

$$= \frac{\text{price at January 1975}}{\text{price at January 1970}} \times 100$$

$$= \frac{60}{40} \times 100 = 150\%$$

The percentage sign is usually omitted and we say that the index is 150 based on January 1970 which is 100. This indicates that the price of the commodity has increased by 50% between January 1970 and January 1975.

EXAMPLE 2. Calculate the price relative for the commodity of Example 1, using January 1975 as the base period.

$$\text{Price relative} = \frac{40}{60} \times 100 = 66.7\%$$

The index is 66.7 based on January 1975 which is 100. This indicates that the price of the commodity in January 1970 was 66.7% of the price in January 1975 or, alternatively the price of the commodity was 33.3% less in January 1970.

EXAMPLE 3. The profits of the MA Company for 6 successive years are given in the table below.

(a) Using Year 1 as a base find an index number of profit.

(b) Using Year 4 as a base calculate the index numbers for the other years.

Year	1	2	3	4	5	6
Profit (£M)	1.3	1.5	1.8	2.3	2.4	2.7

(a) Using Year 1 as the base year we have:

$$\text{Index for Year 2} = \frac{1.5}{1.3} \times 100 = 115.$$

$$\text{Index for Year 3} = \frac{1.8}{1.3} \times 100 = 138 \text{ and so on.}$$

The table below shows the index numbers for all the years in question:

Year	1	2	3	4	5	6
Index	100	115	138	177	185	208

(b) Using Year 4 as base, we have

$$\text{Index for Year 1} = \frac{1.3}{2.3} \times 100 = 57.$$

$$\text{Index for Year 6} = \frac{2.7}{2.3} \times 100 = 117, \text{ and so on.}$$

The table below shows the index numbers for all the years in question:

Year	1	2	3	4	5	6
Index	57	65	78	100	104	117

EXAMPLE 4. The table below shows the index of a certain commodity for the period 1975 − 80. (1960 = 100). Find the index numbers taking 1975 as the base year.

Year	1975	1976	1977	1978	1979	1980
Index	163	175	205	227	254	304

$$\text{Index for 1976} = \frac{175}{163} \times 100 = 107$$

$$\text{Index for 1979} = \frac{254}{163} \times 100 = 156, \text{ and so on.}$$

The full series of index numbers is given below:

Year	1975	1976	1977	1978	1979	1980
Index	100	107	126	139	156	187

10.3 QUANTITY OR VOLUME RELATIVES

Instead of comparing the price of a commodity we can compare the quantity or volume produced or the quantity consumed, etc. In these cases we use the term *quantity* or *volume relatives*.

EXAMPLE 5. The quantity of grain produced by a certain farm in 1974 was 1800 tonnes. In 1980 the quantity produced was 2300 tonnes. Using 1974 as the base year calculate the quantity relative for 1980

$$\text{Quantity relative for 1980} = \frac{\text{quantity produced in 1980}}{\text{quantity produced in 1974}} \times 100$$

$$= \frac{2300}{1800} \times 100 = 128$$

The quantity relative in 1980 is 128 based on 1974 which is 100. Thus 28% more grain was produced in 1980 than in 1974.

10.4 VALUE RELATIVES

If p is the price of a commodity and q is the quantity sold then the value of the commodity is pq. Thus if the price of butter is £1.20 per kilogram and 100 kilograms are sold then the total value of the butter sold is

$$\text{£1.20} \times 100 = \text{£120}$$

If p_o and q_o represent the price and quantity sold during the base period and p_n and q_n are the price and quantity sold during a given period then

$$\text{Value relative} = \frac{p_n q_n}{p_o q_o} \times 100$$

EXAMPLE 6. The price of an article in 1976 was £20 and 2000 were sold. In 1979 the price was £30 and 1500 were sold. Calculate the value relative (1976 = 100).

$$\text{Value relative} = \frac{30 \times 1500}{20 \times 2000} \times 100 = 112.5$$

EXAMPLE 7. A company expects its sales to increase by 60% during the next year. Find the percentage increase in price that will be needed in order that its gross income be doubled.

Price relative $\times$ quantity relative $=$ value relative $\times 100$

Price relative $\times 160 = 200 \times 100$

$$\text{Price relative} = \frac{200 \times 100}{160} = 125$$

Hence the price should be increased by 25%.

10.5 INDEX NUMBER AS A WEIGHTED AVERAGE

An index can be produced which is the average of a series of price relatives. For the index to be realistic it should take into account the amount of each commodity used. The method of weighting allows this to be done.

EXAMPLE 8. Using 1978 as the base year, calculate the weighted arithmetic mean of the price relatives shown in the following table.

Commodity	Quantity used (kilograms)	Price (pence per kg)	
		1978	1979
Margarine	2	70	75
Butter	1	110	120
Cheese	3	90	95

The first step is to calculate the price relatives for the three commodities using 1978 as base. The calculation of the weighted arithmetic mean is then as follows:

Commodity	Price relatives	Weight	Price relative $\times$ weight
Margarine	107	2	214
Butter	109	1	109
Cheese	106	3	318

$$\text{Mean price relative} = \frac{\text{sum of the price relative} \times \text{weight}}{\text{sum of the weights}}$$

$$= \frac{214 + 109 + 318}{2 + 1 + 3} = \frac{641}{6} = 106.8$$

EXAMPLE 9. In order to determine a 'cost of living index' each of six items was given a weight and an initial price relative to 100 at the beginning of a certain year. During the year the increases in price occurred as shown in the table below. Determine the 'cost of living' index at the beginning and end of the year.

Item	Weight	% increase in price
Food	300	30
Household goods	120	20
Clothing	90	15
Housing	90	10
Fuel and light	70	22
Miscellaneous	100	14

Item	Weight	Price relative	Weight × price relative
Food	300	130	39 000
Household goods	120	120	14 400
Clothing	90	115	10 350
Housing	90	110	9 900
Fuel and light	70	122	8 540
Miscellaneous	100	114	11 400
	770		93 590

Cost of living index at beginning of the year = 100

Cost of living index at end of the year = $\dfrac{93\,590}{770}$ = 121.5

10.6 CHAIN BASE INDEXES

Index numbers are sometimes calculated by the chain base method when each year's index number uses the preceding year as a base.

EXAMPLE 10. The prices of a commodity in the years 1974 − 79 are given in the table below.

Year	1974	1975	1976	1977	1978	1979
Price (pence per kg)	30	34	32	36	45	52

Calculate the index numbers for each year using the chain base method. (1974 = 100).

Year	Index No.
1974	100
1975	$\frac{34}{30} \times 100 = 113.3$
1976	$\frac{32}{34} \times 100 = 94.1$

Year	Index No.
1977	$\frac{36}{32} \times 100 = 112.5$
1978	$\frac{45}{36} \times 100 = 125.0$
1979	$\frac{52}{45} \times 100 = 115.6$

Chain base index numbers can also be weighted as shown in Example 11.

EXAMPLE 11. Given the information in the table below calculate the index numbers for each year using the chain base method.

Commodity	Weighting	Price (pence)		
		1977	1978	1979
X	2	12	15	17
Y	5	18	22	30
Z	3	10	13	15

1977 Commodity	Price (pence)	Index No.	Weighting	Index No. × weighting
X	12	100	2	200
Y	18	100	5	500
Z	10	100	3	300
			10	1000

Chain base index number $= \dfrac{1000}{10} = 100$

1978 Commodity	Price (pence)	Chain base index No.	Weighting	Index No. X weighting
X	15	$\dfrac{15}{12} \times 100 = 125$	2	250
Y	22	$\dfrac{22}{18} \times 100 = 122$	5	610
Z	13	$\dfrac{13}{10} \times 100 = 130$	3	390
			10	1250

Chain base index number $= \dfrac{1250}{100} = 125$

1979 Commodity	Price (pence)	Chain base index No.	Weighting	Index No. X weighting
X	17	$\dfrac{17}{15} \times 100 = 113$	2	226
Y	30	$\dfrac{30}{22} \times 100 = 136$	5	680
Z	15	$\dfrac{15}{13} \times 100 = 115$	3	345
			10	1251

Chain base index number $= \dfrac{1251}{10} = 125.1$

It is sometimes necessary to convert chain base index numbers back to find base index numbers. The method of doing this is shown in Example 12.

EXAMPLE 12. The index numbers below have been calculated by the chain base method. Convert these to a fixed base (1975 = 100).

Year	1975	1976	1977	1978	1979
Index No.	100	150	130	160	180

Solution

Year	Index No. based on 1975
1976	150
1977	$\dfrac{130 \times 150}{100} = 195$
1978	$\dfrac{160 \times 130 \times 150}{100 \times 100} = 312$
1979	$\dfrac{180 \times 160 \times 130 \times 150}{100 \times 100 \times 100} = 561.6$

The fixed base index numbers obtained may easily be checked by the method shown below:

Year	Chain base index number
1977	$\dfrac{195}{150} \times 100 = 130$
1978	$\dfrac{312}{195} \times 100 = 160$
1979	$\dfrac{561.6}{312} \times 100 = 180$

When weighted index numbers have been used the conversion from chain based to fixed base cannot be done accurately although an approximation can be made.

10.7 COMPARISON OF FIXED BASE AND CHAIN BASE INDEXES

The fixed base method has the advantages that it is easy to calculate and easy to understand and years far apart can be quickly compared. The disadvantages are that conditions tend to change over a period and after a period of time the fixed base index number becomes unrealistic.

The chain-based method has the advantages that the index numbers are always up to date and that revised weightings can be introduced at any time. Also, new items can be included and original ones removed at any time. The disadvantages are that it is difficult to compare the index numbers over a period of years.

EXERCISE 16

1. In April, 1970 the price of a commodity was 90 pence whilst in April, 1980 its price, for the same quantity, was 120 pence. Calculate the price relative:
    (a) taking 1970 as base   (b) taking 1980 as base.

2. The table below shows the price per tin of a brand of instant coffee in the years 1975 – 80 inclusive. Calculate the price relative for each year:

   (a) using 1975 as base   (b) using 1978 as base.

Year	1975	1976	1977	1978	1979	1980
Price (pence)	60	70	84	100	120	144

3. The table below shows the index of steel prices for the period 1976 – 80 inclusive (1970 = 100). Calculate the index numbers using 1976 as base.

Year	1976	1977	1978	1979	1980
Index	108	110	115	120	128

4. In 1970 a steel stockist sold 5 million tonnes of steel products. In 1980 sales were 8 million tonnes. Calculate the quantity relative:

   (a) using 1970 as base   (b) using 1980 as base.

5. The table below gives the number of television licences in the UK. Using 1955 as base calculate the quantity relatives for the other years.

Year	1955	1960	1965	1970
Number (millions)	4.0	10.5	14.2	17.1

6. The quantity relative for 1980 with 1960 as base was 105 whilst the quantity relative for 1980 with 1970 as base was 140. Find the quantity relative for 1960 with 1970 as base.

7. The price relative for a commodity is 127 and the quantity relative is 110. What is the value relative?

8. A company expects the sales of its product to increase by 80% in the coming year. If its gross income is to be increased by 250% by what percentage should it increase its selling price?

9. The table below gives the price of certain commodities in the years 1978 and 1979. Using 1978 as base calculate the arithmetic mean of the price relatives.

Commodity	Price per (kg)	
	1978	1979
Potatoes	4	7
Carrots	6	9
Swedes	3	5

10. Calculate the index based upon a simple aggregate of the prices given in question 9, using 1979 as base.

11. The table below shows the retail prices of clothing and footwear. Calculate an index number for the retail prices of clothing and footwear.

Item	Weight	Indices 15-12-1980 (16-1-1972 = 100)
Men's outer clothing	14	137
Men's underclothing	5	134
Women's outer clothing	20	125
Women's underclothing	5	127
Children's clothing	11	126
Other clothing	14	119
Footwear	17	130

12. The following table shows the US consumption (millions of pounds) and price (dollars per pound) of vegetable oil products.

Type of oil	1973		1974		1975	
	Quantity	Price	Quantity	Price	Quantity	Price
Soybean	322	0.13	368	0.12	367	0.13
Cotton seed	96	0.15	114	0.14	123	0.14
Linseed	32	0.13	31	0.13	19	0.12

Taking 1973 as the base year, calculate index numbers for the general level of prices of these products for 1974 and 1975.

13. The following table shows the average wholesale price and production quantities of various commodities supplied by a local firm for the years 1970 and 1980. Compare a weighted price index for the year 1980 using 1970 as base.

Item	Price per kilo (£)		
	1970	1980	Weights
Copper brads	3.89	4.13	5
Bronze clips	62.2	59.7	3
Brass studs	35.4	38.9	2

14. The prices of a commodity in the years 1974 – 79 are given in the table below. Calculate the index numbers for each year using the chain base method with 1974 = 100.

Year	1974	1975	1976	1977	1978	1979
Price (£ per tonne)	28	29	32	38	43	50

15. Using the chain base method with 1977 = 100, calculate the index numbers for 1978 and 1979 for the data given below.

Commodity	Weighting	Price (£)		
		1977	1978	1979
A	2	2.00	2.50	3.00
B	3	3.10	3.00	3.20
C	5	4.50	4.60	5.00

16. Given the information in the table below calculate the index numbers for each year using the chain base method with 1975 = 100.

Item	Weighting	Price (£)		
		1975	1976	1977
A	3	5	7	8
B	4	7	8	10
C	6	10	9	11

17. The index numbers shown below have been calculated by the chain base method. Convert them to a fixed base (1976 = 100).

Year	1976	1977	1978	1979	1980
Index number	100	110	115	118	122

11 ANALYSIS OF TIME SERIES

11.1 INTRODUCTION

A time series is a set of observations taken at specified times, usually at equal intervals, i.e. daily, monthly, yearly, etc. Some examples are:

(i)   total annual production of steel in the UK
(ii) monthly sales of cars in the EEC
(iii) quarterly values of imports into the UK.

11.2 GRAPHS OF TIME SERIES

A time series is represented pictorially by drawing a graph, which is sometimes called a *historigram*. The graph of a time series showing the total expenditure on highways in ten successive years is shown in Fig. 11.1.

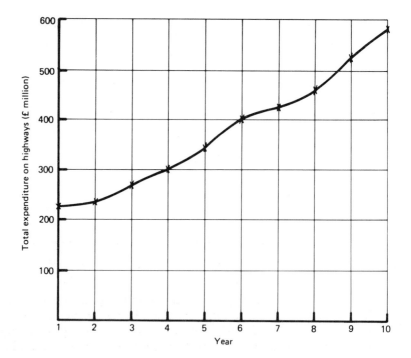

Fig. 11.1.

11.3 TIME SERIES MOVEMENTS

A time series is subject to certain variations or movements. The analysis of these movements assists in forecasting future movements. The characteristic movements of a time series may be classified as follows:

1) *Long-term or secular movements.* These refer to the trend of the graph of a series over a long period of time. The long-term movement may be represented by a trend curve (Fig. 11.2) or a trend line (Fig. 11.3). A typical long-term trend is the growth of a company over a period of years.

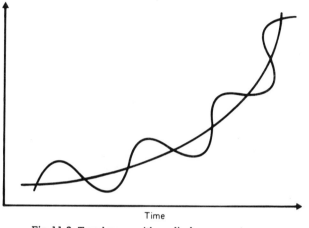

Time

Fig. 11.2. Trend curve with cyclical movement

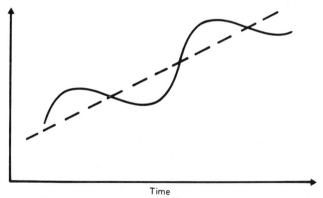

Time

Fig. 11.3. Long term trend and cyclical movement

2) *Cyclical movements* which refer to swings about the trend line or curve. These cylces are often periodic, that is they follow similar patterns after equal periods of time. In the business world movements are only considered to be cyclical if they recur at intervals of more than a year. A typical cyclical movement is the business cycle representing periods of prosperity, recession, slump and recovery. In Figs. 11.2 and 11.3, the cyclical movements are clearly shown.

3) *Seasonal movements* occur because of events which take place each year. An example is the increase in the number of people employed in agriculture during the fruit and potato harvesting period. In Fig. 11.1 no seasonal variations occur because annual figures for expenditure were taken. However, the graph of Fig. 11.4 shows the seasonal variations in the sales of a store. For instance, the sales in the fourth quarter are always high because of Christmas followed by a slump in sales during the first quarter of the following year (after the January sales of course). Although in the business world, seasonal movements refer to annual variations the ideas can be extended to any period of time, i.e. hourly, daily, weekly, etc., depending upon the information available.

4) *Irregular movements* which occur because of chance events like strikes, floods, etc. Usually the events producing irregular events last only a short time but sometimes they can produce new cyclical variations.

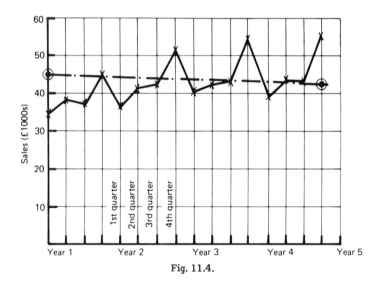

Fig. 11.4.

11.4   ANALYSIS OF TIME SERIES

Analysing a time series consists of splitting it up into its component movements. Fig. 11.5 (a) shows a long-term trend line whilst (b) shows a cyclical variation imposed upon it. Irregular movements have been superimposed in (c) and the result looks like a time series occurring in practice.

A time series may be represented by the equation

$$Y = T + C + S + I$$

where T is the trend

C is the cyclical movement

S is the seasonal movement

and I is the irregular movement.

A First Course in Statistics

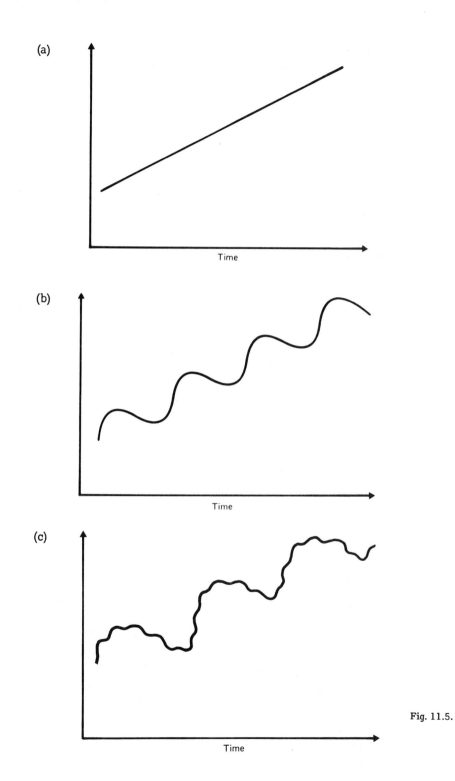

(a)

Time

(b)

Time

(c)

Time

Fig. 11.5.

11.5 TREND ANALYSIS

Since the trend is concerned with long-term movements, the trend is usually found by using annual data. Ideally, data from at least fifteen years should be used so that cyclical movements extending over several years are not taken to indicate an overall trend of the time series.

The method of drawing the best straight line (or the best curve) is often used for finding the trend component of a time series.

As shown in Chapter 7, the linear equation is

$$y = a + bx$$

and this equation may be used if the trend seems to follow a straight line.

EXAMPLE 1. The table below shows the number of cars on the road in the UK during a 15 year period. Calculate the trend.

Year	Number of cars (100 000s)
1	32
2	38
3	45
4	53
5	60
6	66
7	74
8	82
9	89
10	95
11	103
12	108
13	112
14	115
15	121

The graph of the time series is drawn in Fig. 11.6. Note that time is always taken as being the independent variable. As before, to draw the best straight line we find $\bar{x}$ and $\bar{y}$. Since there are 15 years, $\bar{x}$ = 8.0 years. The total number of cars on the road is 1193 hundred thousands. Hence $\bar{y}$ = $\dfrac{1193}{15}$ = 79.5.

By drawing the right-angled triangle ABC,

$$\text{gradient} = a = \frac{66}{10} = 6.6$$

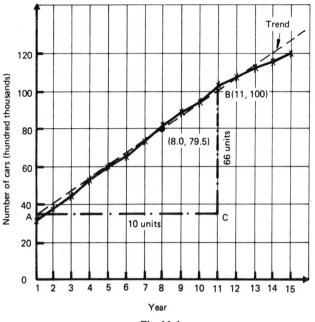

Fig. 11.6.

To find the value of the intercept on the y-axis we use the coordinates of the point B(11, 100).

$$100 = a + 6.6 \times 11$$

$$a = 27$$

Hence the equation of the trend line is $y = 27 + 6.6x$.

To forecast the number of cars that will be on the road in Year 16, we substitute for $x = 16$, giving

$$y = 27 + 6.6 \times 16 = 132.6$$

Hence we forecast that there will be 13 200 000 cars on the road in Year 16.

11.6 METHOD OF MOVING AVERAGES

Moving averages tend to reduce the amount of variation present in a time series. Thus the cyclical, seasonal and irregular movements present in the series may be eliminated thereby leaving only the trend.

Given a set of numbers $Y_1, Y_2, Y_3, \ldots$ a moving average of order N is given by the following sequence of means:

$$\frac{Y_1 + Y_2 + \ldots + Y_N}{N}, \ \frac{Y_2 + Y_3 + \ldots + Y_{N+1}}{N}, \ \frac{Y_3 + Y_4 + \ldots + Y_{N+2}}{N}, \ldots$$

The sums of the numerators are called *moving totals*.

EXAMPLE 2. Given the set of numbers 3, 5, 8, 4, 7, 3, 6, obtain a moving average of order 4.

The mean of the first four numbers is:

$$\frac{3+5+8+4}{4} = 5$$

To obtain the next mean, remove the first number which is 3 and replace it by the fifth number which is 7. Thus

$$\frac{5+8+4+7}{4} = 6$$

The next mean is obtained by removing the number 5 and replacing it by the sixth number which is 3.

$$\frac{8+4+7+3}{4} = 5.5$$

The last mean is obtained from

$$\frac{4+7+3+6}{4} = 5$$

Hence the moving average of order 4 is 5, 6, 5.5 and 5.

If the data are given weekly we say that a moving average of order N is an N weekly moving average. Thus we speak of a 3 year moving average or a 12 month moving average.

EXAMPLE 3. The figures below show the sales, in thousands, of a manufacturing company during three successive years.

	Year 1	Year 2	Year 3
Quarter 1	20.0	21.8	23.1
Quarter 2	18.1	19.4	21.3
Quarter 3	15.2	16.6	18.0
Quarter 4	11.0	12.4	14.2

(a) Calculate the four quarterly moving averages.

(b) Represent the above figures on a graph and mark on it the four quarterly moving averages.

(c) Draw the trend line on the graph and use it to forecast sales for the first and second quarters of Year 4.

		Y	Moving total	Moving average
Year 1	1	20.0		
	2	18.1		
			64.3	16.1
	3	15.2		
			66.1	16.5
	4	11.0		
			67.4	16.9
Year 2	1	21.8		
			68.8	17.2
	2	19.4		
			70.2	17.6
	3	16.6		
			71.5	17.9
	4	12.4		
			73.4	18.4
Year 3	1	23.1		
			74.8	18.7
	2	21.3		
			76.6	19.2
	3	18.0		
	4	14.2		

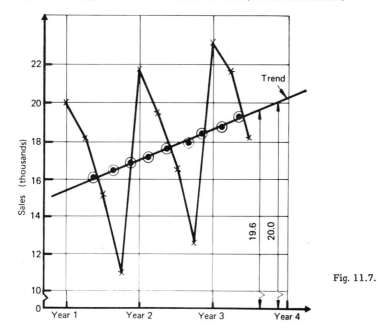

Fig. 11.7.

The graph is shown in Fig. 11.7. The moving averages have been plotted half-way between quarters because half-way for any four values takes us to a point mid-way between the second and third values. The moving averages deviate only slightly from a straight line and this 'best straight line' is taken as being the trend.

To forecast sales for the first quarter of Year 4 we use the trend to obtain the value of the moving average corresponding to Year 3, quarter 3½. From the graph this value is 19.6.

Let the sales for Year 4, Quarter 1, be x, then

$$\frac{21.3 + 18.0 + 14.2 + x}{4} = 19.6$$

$$53.5 + x = 78.4$$

$$x = 24.9$$

From the graph the value of the moving average corresponding to Year 3, quarter 4½ is 20.0.

Let the sales for Year 4, Quarter 2, be y, then

$$\frac{18.0 + 14.2 + 24.9 + y}{4} = 20.0$$

$$57.1 + y = 80.0$$

$$y = 22.9$$

One disadvantage of the moving average method is that data at the beginning and end of the series are lost. Thus, in Example 3, the data for Year 1, Quarters 1 and 2 and for Year 3, Quarters 3 and 4 are lost.

It is very important to decide on the order of the moving average. This depends upon the original values in the series. For instance, if the pattern repeats itself every third value then we would use a three-point moving average. In Example 3 we see that the pattern repeats itself every fourth value and for this reason a four-point moving average has been used. Sometimes the trend approximates to a smooth curve. In such cases advanced methods, beyond the scope of this book, have to be used in order to make forecasts.

EXAMPLE 4. The information below relates to the sales of women's shoes (in millions of pairs).

	Year 1	Year 2	Year 3
1st quarter	20.9	17.5	17.0
2nd quarter	17.3	14.7	13.5
3rd quarter	15.6	13.5	13.5
4th quarter	13.9	13.1	13.7

Calculate the four quarterly moving average and draw a graph of the original information with the trend superimposed upon it.

A First Course in Statistics

		Y	Moving total	Moving average
Year 1	1	20.9		
	2	17.3		
			67.7	16.9
	3	15.6		
			64.3	16.1
	4	13.9		
			61.7	15.4
Year 2	1	17.5		
			59.6	14.9
	2	14.7		
			58.8	14.7
	3	13.5		
			58.3	14.6
	4	13.1		
			57.1	14.3
Year 3	1	17.0		
			57.1	14.3
	2	13.5		
			57.7	14.4
	3	13.5		
	4	13.7		

The graph is shown in Fig. 11.8, and it will be seen that the trend approximates to a smooth curve.

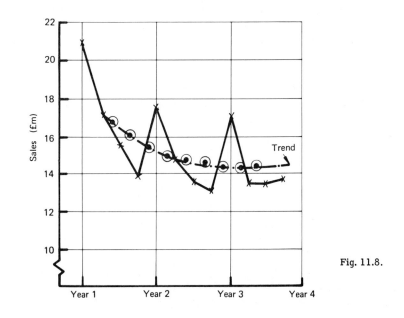

Fig. 11.8.

1. Find the moving averages of order 2 for the following set of numbers:

    2, 6, 4, 3, 1, 2.

2. Find the moving averages of order 3 for the set of numbers given in question 1.

3. Construct a four quarterly moving average and hence find the trend for the data given below. Graph the trend line and the original data.

    Live births in the UK (in thousands)

	Quarter 1	Quarter 2	Quarter 3	Quarter 4
Year 1			237.3	221.6
Year 2	248.7	240.0	239.8	220.7
Year 3	244.8	237.6	229.4	209.1
Year 4	225.6	235.1		

4. By drawing the best straight line obtain the trend line for the data given below and hence forecast the production of paper sacks in Year 12.

    Annual production of paper sacks (thousands)

Year	Production
1	50.0
2	36.5
3	43.0
4	44.5
5	38.9
6	38.1
7	32.6
8	38.7
9	41.7
10	41.1
11	33.8

5. By drawing the best straight line determine the equation for the trend line assuming that this is linear. Use your trend line to forecast the profits in Year 16.

Profits of a company (£thousands)

Year	Profit
1	65
2	77
3	89
4	105
5	120
6	133
7	150
8	165
9	180
10	190
11	205
12	215
13	225
14	231
15	243

6. The following data relate to the production (in thousands of tonnes) of compound feeding stuffs. By calculating a four-quarterly moving average obtain the trend. Draw a graph of the original data and superimpose on it the points representing the moving averages.

	Quarter 1	Quarter 2	Quarter 3	Quarter 4
Year 1		800	800	1010
Year 2	1025	880	810	970
Year 3	960	770	750	910
Year 4	880			

7. The figures below show the quarterly sales of a small company in £thousands.

	Quarter 1	Quarter 2	Quarter 3	Quarter 4
Year 1	81	46	41	76
Year 2	82	47	41	77
Year 3	83	48	42	77
Year 4	84	49		—

(a) Use a four quarterly moving average to calculate the trend.
(b) Draw a graph of the original data and superimpose on it the best straight line which represents the trend.
(c) Use the trend to forecast the sales in Quarters 3 and 4 of Year 4.

12 PROBABILITY

12.1 SIMPLE PROBABILITY

If a fair coin is tossed the outcome is equally likely to be heads or tails. We say that the events of tossing heads or tossing tails are *equi-probable events*.

Similarly if we throw an unbiased die (plural: dice) the events of throwing a 1, or a 2, or a 3, or a 4, or a 5, or a 6 are equi-probable events.

The probability of an event occurring is defined as

$$P(E) = \frac{\text{total number of favourable outcomes}}{\text{total number of possible outcomes}}$$

EXAMPLE 1. (a) A fair coin is tossed once. What is the probability that it will come down heads?

Number of favourable outcomes = 1; total number of possible outcomes = 2.

$$P(\text{head}) = \frac{1}{2}$$

(b) What is the probability of scoring a 5 in the single roll of a fair die?

Number of favourable outcomes = 1; total number of possible outcomes = 6.

$$P(\text{five}) = \frac{1}{6}$$

(c) What is the probability of drawing an Ace from a deck of 52 playing cards when the deck is cut once?

Number of favourable outcomes = 4 (since there are four Aces in the deck); total number of possible outcomes = 52.

$$P(\text{Ace}) = \frac{4}{52} = \frac{1}{13}$$

12.2 PROBABILITY SCALE

When an event is absolutely certain to happen we say that the probability of it happening is 1. Thus the probability that one day each of us will die is 1. When an event can never happen we say the probability of it happening is 0. Thus the probability that any one of us can jump a height of 5 metres unaided is 0.

All probabilities must therefore have a value between 0 and 1. They can be expressed as either a fraction or a decimal. Thus

$$P \text{ (head)} = \frac{1}{2} = 0.5$$

$$P \text{ (Ace)} = \frac{1}{13} = 0.077$$

Probabilities can be expressed on a probability scale (Fig. 12.1).

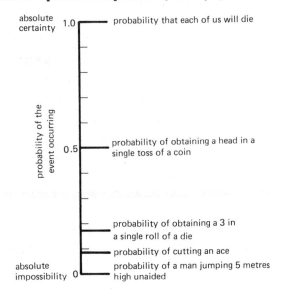

Fig. 12.1.

12.3 TOTAL PROBABILITY

If we toss a fair coin it will come down either heads or tails. That is:

$$P \text{ (heads)} = \frac{1}{2} \text{ and } P \text{ (tails)} = \frac{1}{2}$$

The total probability covering all possible outcomes is ½ + ½ = 1. Another way of saying this is

$$P \text{ (favourable outcomes)} + P \text{ (unfavourable outcomes)} = 1$$

EXAMPLE 2. A bag contains 5 blue balls, 3 red balls and 2 black balls. A ball is drawn at random from the bag. Calculate the probability that it will be (a) blue, (b) red, (c) not black.

(a) $P \text{ (blue)} = \frac{5}{10} = 0.5$

(b) $P \text{ (red)} = \frac{3}{10} = 0.3$

(c) P (black) $= \dfrac{2}{10} = 0.2$

P (not black) $= 1 - 0.2 = 0.8$

12.4 MEANING OF PROBABILITY

When we say that the probability of an event happening is $1/3$ we do *not* mean that if we repeat the experiment three times the event will happen once. Even if we repeat the experiment 30 times it is very unlikely that the event will happen exactly 10 times.

Probability tells us what to expect in the long run. If the experiment is repeated 300 times we would expect the event to happen *about* 100 times. However we would not be worried if it happened only on 94 occasions or if it happened on 106 of them.

In cases where the probability has to be determined by experiment, the probability of an event happening is calculated from the formula:

$$P(E) = \frac{\text{number of trials with favourable outcomes}}{\text{total number of trials}}$$

Probabilities determined by experiment are often called *empirical probabilities*.

EXAMPLE 3. 50 invoices are checked for errors. If 3 of them are found to contain errors, determine the probability that an invoice, chosen at random from these 50, will contain errors.

Treating the event of finding an error as a favourable outcome:

number of favourable outcomes $= 3$; total number of trials $= 50$.

$$P \text{ (finding an error)} = \frac{3}{50} = 0.06$$

12.5 RELATIVE FREQUENCY AND PROBABILITY

The relative frequency of a class in a frequency distribution is found by dividing the class frequency by the total frequency. That is,

$$\text{Relative frequency} = \frac{\text{class frequency}}{\text{total frequency}}$$

EXAMPLE 4. A loaded die was thrown 200 times with the following results:

Score	1	2	3	4	5	6
Frequency	28	36	32	30	34	40

Calculate the relative frequencies.

The relative frequencies are found by dividing each of the class frequencies by 200. (Note that the sum of the relative frequencies must equal 1.)

Score	1	2	3	4	5	6
Relative frequency	0.14	0.18	0.16	0.15	0.17	0.20

In the case of the loaded die (Example 4), the probabilities of obtaining each of the scores on the die can only be determined by experiment. We can estimate the probabilities by assuming that they are the same as the relative frequencies. Certainly, if the number of observations is very large the relative frequencies will give results which are very near the truth.

Consider the following as an example. If 1000 tosses of a coin result in 542 heads, the relative frequency of heads is 542/1000 = 0.542. If in another 1000 tosses the number of heads is 492, the total number of heads in 2000 tosses is 542 + 492 = 1034. The relative frequency is then 1034/2000 = 0.517. By continuing in this way we should get closer and closer to 0.5, which is the calculated probability of throwing a head in a single toss of the coin.

EXERCISE 18

1. A die is rolled. Calculate the probability that it will give:

    (a) a five, (b) a score less than 3, (c) an even number.

2. A card is drawn at random from a pack of 52 playing cards. Calculate the probability that it will be (a) the Jack of Hearts, (b) a King, (c) an Ace, King, Queen, or Jack, (d) the King of Hearts or the Ace of Spades.

3. A letter is chosen from the word TERRIFIC. Determine the probability that it will be (a) an R, (b) a vowel, (c) a consonant.

4. A bag contains three red balls, five blue balls and two green balls. A ball is chosen at random from the bag. Calculate the probability that it will be (a) green, (b) blue, (c) not red.

5. Two dice are thrown and their scores are added together. Find the probability that the total will be (a) 5, (b) less than 5, (c) more than 5.

6. Determine the probability for each of the following situations:

(a) A sample of 9000 industrial workers were questioned regarding industrial injuries. 600 reported sustaining such injuries during a 12 month period. What is the probability of a worker sustaining an industrial injury?

(b) A wholesaler of electrical goods finds that of 150 deliveries from a certain firm, 10 are late. Calculate the probability of the next delivery being late.

(c) A new component is fitted to an engine. 20 engines, fitted with the new part are tested and 2 fail to function correctly. Calculate the probability that an engine fitted with the new component will not function correctly.

7. The table below shows the family income of 500 families.

Income range (£)	Number of families
Less than 3000	70
3000 and less than 4000	120
4000 and less than 5000	180
5000 and less than 6000	80
More than 6000	50

If a family is chosen at random from these 500 families, calculate the probability that its income will be (a) less than £3000, (b) £4000 but less than £5000, (c) more than £6000.

8. Over a period of 100 days the following numbers of absentees were recorded:

Number of absentees	0	1	2	3
Number of days	14	28	28	18

(a) What is the probability that on any one day there will be no absentees?

(b) On any one day what is the probability that there will be more than 3 absentees?

(c) On any one day what is the probability that there will be less than 3 absentees?

12.6 INDEPENDENT EVENTS

An independent event is one which has no effect on subsequent events. If a die is rolled twice what happens on the first roll does not affect what happens on the second or third rolls. Hence the three rolls of the die are independent events. Similarly the events of tossing a coin then cutting a deck of playing cards are independent events because the way in which the coin lands has no effect on the cut.

If $P(E_1)$ = the probability of the first event happening

$P(E_2)$ = the probability of the second event happening

and $P(E_1 E_2)$ = the probability that *both* E_1 and E_2 occur then $P(E_1 E_2)$ = $P(E_1) \times P(E_2)$

This is known as the multiplication law of probability. In general,

$$P(E_1 E_2 \ldots E_n) = P(E_1) \times P(E_2) \times \ldots \times P(E_n)$$

EXAMPLE 5. A fair coin is tossed and a card is then drawn from a pack of 52 playing cards. Find the probability that a head and an Ace will result.

Let $P(E_1)$ = the probability of obtaining a head

Then $P(E_1)$ = $\dfrac{1}{2}$

Let $P(E_2)$ = the probability of cutting an Ace

Then $P(E_2)$ = $\dfrac{4}{52}$ = $\dfrac{1}{13}$

Let $P(E_1 E_2)$ = the probability that both E_1 and E_2 will occur

Then $P(E_1 E_2)$ = $\dfrac{1}{2} \times \dfrac{1}{13}$ = $\dfrac{1}{26}$

EXAMPLE 6. A coin is tossed five times. What is the probability of each toss resulting in a head?

The five tosses of the coin are independent events since what happens on the first toss in no way affects subsequent tosses. Similarly what happens on the second toss in no way affects the third and subsequent tosses and so on.

$$P(\text{5 heads}) = \dfrac{1}{2} \times \dfrac{1}{2} \times \dfrac{1}{2} \times \dfrac{1}{2} \times \dfrac{1}{2} = \dfrac{1}{32}$$

A Venn diagram (Fig. 12.2) may be used to illustrate the probability of two events E_1 and E_2 occurring. The shaded area gives the required probability.

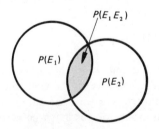

Fig. 12.2.

12.7 DEPENDENT EVENTS

Consider a bag containing 3 red balls and 2 blue balls. A ball is drawn at random from the bag and *not* replaced. The probability that it is red is 3/5. Now let us choose a second ball. The probability that this is also red is 2/4. Hence the probability of drawing two red balls is $3/5 \times 2/4 = 3/10$. The events of drawing one red ball followed by drawing a second red ball are *dependent events* because the probability of the second event depends upon what happened in the first event.

EXAMPLE 7. A bag contains 5 white balls, 3 black balls and 2 green balls. A ball is chosen at random from the bag and not replaced. In three draws find the probability of obtaining white, black and green in that order.

Let $P(E_1)$ = the probability of drawing a white ball on the first draw = $\dfrac{5}{10}$

$P(E_2)$ = the probability of drawing a black ball on the second draw = $\dfrac{3}{9}$

$P(E_3)$ = the probability of drawing a green ball on the third draw = $\dfrac{2}{8}$

The probability of drawing white/black then green

$$= \frac{5}{10} \times \frac{3}{9} \times \frac{2}{8} = \frac{1}{24}$$

12.8 MUTUALLY EXCLUSIVE EVENTS

If two events could not happen at the same time they are said to be *mutually exclusive*. For instance, suppose we want to know the probability of a 3 or a 4 occurring in the single roll of a die. In a single roll either a 3 can occur or a 4 can occur. It is not possible for a 3 and a 4 to occur together. Hence the events of throwing a 3 or a 4 in a single roll of a die are mutually exclusive. Similarly, it is impossible to cut a Jack and a King in a single cut of a pack of playing cards. Hence these two events are mutually exclusive.

If $E_1, E_2, \ldots, E_n$ are mutually exclusive events then the probability of *one* of the events occurring is

$$P(E_1 + E_2 + \ldots + E_n) = P(E_1) + P(E_2) + \ldots + P(E_n)$$

EXAMPLE 8. A die with faces numbered 1 to 6 is rolled once. What is the probability of obtaining either a 3 or a 4?

$$P(3) = P(E_1) = \frac{1}{6}$$

$$P(4) = P(E_2) = \frac{1}{6}$$

$$P(3 \text{ or } 4) = P(E_1 + E_2) = \frac{1}{6} + \frac{1}{6} = \frac{1}{3}$$

EXAMPLE 9. A pack of playing cards is cut once. Find the probability that the card which is cut will be the Ace of Spades, a King, or the Queen of Hearts.

$$P(\text{Ace of Spades}) = P(E_1) = \frac{1}{52}$$

$$P(\text{King}) = P(E_2) = \frac{4}{52}$$

$$P(\text{Queen of Hearts}) = P(E_3) = \frac{1}{52}$$

$$P(\text{Ace of Spades, a King or the Queen of Hearts}) = P(E_1 + E_2 + E_3)$$

$$= \frac{1}{52} + \frac{4}{52} + \frac{1}{52} = \frac{6}{52} = \frac{3}{26}$$

12.9 NON-MUTUALLY EXCLUSIVE EVENTS

If a pack of playing cards is cut once, the events of drawing a Jack and drawing a Diamond are not mutually exclusive events because the Jack of Diamonds can be cut.

If $P(E_1)$ is the probability that an event E_1 will occur and if $P(E_2)$ is the probability that an event E_2 will occur then the probability that either E_1 or E_2 or both E_1 and E_2 will occur is

$$P(\text{Jack of Diamonds}) = P(E_1) + P(E_2) - P(E_1 E_2)$$

EXAMPLE 10. If E_1 is the event of drawing a Jack and E_2 is the event of drawing a Diamond find the probability of drawing a Jack or a Diamond or both in a single cut of a pack of playing cards.

$$= P(E_1) + P(E_2) - P(E_1 E_2)$$

$$= \frac{4}{52} + \frac{13}{52} - \frac{4}{52} \times \frac{13}{52} = \frac{4}{52} + \frac{13}{52} - \frac{1}{52} = \frac{16}{52} = \frac{4}{13}$$

In the case of three mutually exclusive events:

$$P = P(E_1) + P(E_2) + P(E_3) - P(E_1 E_2) - P(E_1 E_3) - P(E_2 E_3) + P(E_1 E_2 E_3)$$

Note that:

$$P = 1 - P(\overline{E_1}\,\overline{E_2}\,\overline{E_3})$$

where $P(\overline{E_1})$ = the probability of E_1 not occurring = $1 - P(E_1)$ and similarly for the other events.

The difference between mutually exclusive events and non-mutually exclusive events may be illustrated by drawing Venn diagrams as shown in Fig. 12.3.

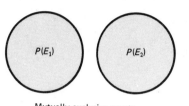

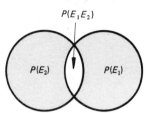

Mutually exclusive events
Shaded area gives
$P(E_1 + E_2) = P(E_1) + P(E_2)$

Non-mutually exclusive events
Shaded area gives
$P(E_1 + E_2) = P(E_1) + P(E_2) - P(E_1E_2)$

Fig. 12.3.

EXAMPLE 11. Three people A, B and C work independently at solving a crossword puzzle. The probability that A will solve the problem is 2/3, the probability that B will solve it is 3/4 and the probability that C will solve it is 4/5. Determine the probability that the puzzle will be solved.

We have $P(E_1) = \dfrac{2}{3}$ $P(E_2) = \dfrac{3}{4}$ and $P(E_3) = \dfrac{4}{5}$

Hence $P(\overline{E_1}) = 1 - P(E_1) = 1 - \dfrac{2}{3} = \dfrac{1}{3}$

$P(\overline{E_2}) = 1 - P(E_2) = 1 - \dfrac{3}{4} = \dfrac{1}{4}$

$P(\overline{E_3}) = 1 - P(E_3) = 1 - \dfrac{4}{5} = \dfrac{1}{5}$

$P = 1 - P(\overline{E_1}\,\overline{E_2}\,\overline{E_3})$

$= 1 - \dfrac{1}{3} \times \dfrac{1}{4} \times \dfrac{1}{5} = 1 - \dfrac{1}{60} = \dfrac{59}{60}$

12.10 PROBABILITY TREE

Suppose that we toss a coin three times. What are the various possibilities and what are their respective probabilities? One way of finding out is to draw a probability tree.

On the first toss the coin can show either a head or a tail. The probability of a head is ½ and the probability of a tail is also ½. Showing possible heads by a full line and possible tails by a dotted line we can draw Fig. 12.4 (a).

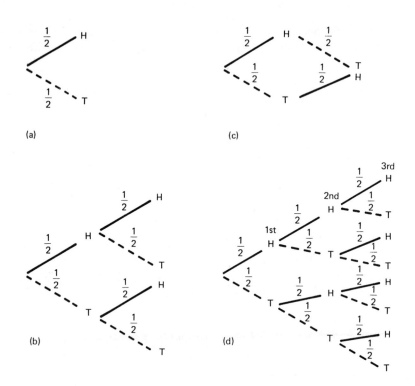

Fig. 12.4.

On the second toss, for each of the branches in Fig. 12.4 (a) we may obtain either a head or a tail. Hence from each of the branches in diagram (a) we draw two more branches as shown in diagram (b). Diagram (b) tells us that the probability of a head occurring in both tosses is

$$P(HH) = \frac{1}{2} \times \frac{1}{2} = \frac{1}{4}$$

One head may be obtained in two ways, i.e. a head followed by a tail or a tail followed by a head as shown in Fig. 12.4 (c). Hence

$$P(\text{one head}) = \frac{1}{2} \times \frac{1}{2} + \frac{1}{2} \times \frac{1}{2} = \frac{1}{4} + \frac{1}{4} = \frac{1}{2}$$

Carrying on the same way the tree diagram is completed for the three tosses of the coin as shown in Fig. 12.4 (d).

EXAMPLE 12. Using the tree diagram (Fig. 12.4):

(a) write down all the possibilities that can occur when the coin is tossed three times

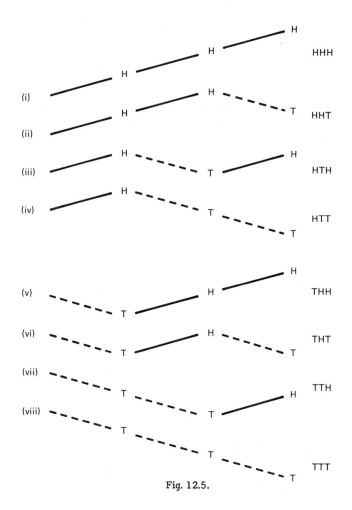

Fig. 12.5.

(b) calculate the probability of three heads occurring

(c) calculate the probability that only one head will appear in the three tosses

(d) calculate the probability of three tails occurring.

(a) The possibilities are as shown in Fig. 12.5.

(b) Using branch (i) of Fig. 12.5 gives us

$$P(\text{HHH}) = P\text{ (three heads)} = \frac{1}{2} \times \frac{1}{2} \times \frac{1}{2} = \frac{1}{8}$$

(c) Using branches (iv), (vi) and (vii) of Fig. 12.5 gives us

$$P(\text{one head}) = \frac{1}{2} \times \frac{1}{2} \times \frac{1}{2} + \frac{1}{2} \times \frac{1}{2} \times \frac{1}{2} + \frac{1}{2} \times \frac{1}{2} \times \frac{1}{2} = \frac{3}{8}$$

(d) Using branch (viii) of Fig. 12.5 gives us

$$P(\text{TTT}) = P(\text{three tails}) = \frac{1}{2} \times \frac{1}{2} \times \frac{1}{2} = \frac{1}{8}$$

EXAMPLE 13. A box contains 4 black and 6 red balls. A ball is drawn from the box and it is not replaced. A second ball is then drawn. Find the probabilities of:

(a) red then black being drawn

(b) black then red being drawn

(c) red then red being drawn

(d) black then black being drawn.

The probability tree and the probabilities are shown in Fig. 12.6.

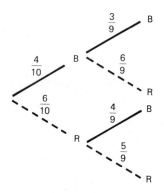

Fig. 12.6.

(a) $P(\text{RB}) = \dfrac{6}{10} \times \dfrac{4}{9} = \dfrac{4}{15}$

(b) $P(\text{BR}) = \dfrac{4}{10} \times \dfrac{6}{9} = \dfrac{4}{15}$

(c) $P(\text{RR}) = \dfrac{6}{10} \times \dfrac{5}{9} = \dfrac{1}{3}$

(d) $P(\text{BB}) = \dfrac{4}{10} \times \dfrac{3}{9} = \dfrac{2}{15}$

12.11 REPEATED TRIALS

If in a single trial, p is the probability of success and q is the probability of failure, then $p + q = 1$, since the trial must result in either success or failure.

If there are n trials then the successive terms of $(q + p)^n$ give the probabilities of 0, 1, 2, ..., n successes.

Pascal's triangle, as shown below, may be used to expand expressions of the type $(q + p)^n$.

Value of n	Numerical coefficients of $(q + p)^n$

```
                                    1
   1                             1     1
   2                          1     2     1
   3                       1     3     3     1
   4                    1     4     6     4     1
   5                 1     5    10    10     5     1
   6              1     6    15    20    15     6     1
   7           1     7    21    35    35    21     7     1
   8        1     8    28    56    70    56    28     8     1
```

Each number in the triangle is obtained from the line immediately above by adding together the two numbers which lie on either side of it. Thus in the line $n = 6$, the number 20 is obtained by adding together the numbers 10 and 10 in the line $n = 5$. Also in the line $n = 8$, the number 28 is obtained by adding the numbers 21 and 7 in the line $n = 7$.

The expansion of $(q + p)^n$ is easily obtained if we remember that the powers of q decrease from q^n (which is always the first term) and that powers of p increase to p^n (which is always the last term). Thus

$$(q + p)^4 = q^4 + 4q^3p + 6q^2p^2 + 4qp^3 + p^4$$
$$(q + p)^6 = q^6 + 6q^5p + 15q^4p^2 + 20q^3p^3 + 15q^2p^4 + 6qp^5 + p^6$$

EXAMPLE 14. A die is rolled four times. Calculate the probabilities of obtaining no sixes, 1 six, 2 sixes, 3 sixes and 4 sixes in the four rolls.

Let the event of rolling a six in a single roll be regarded as a success. Then

$$p = \frac{1}{6}$$

and

$$q = 1 - \frac{1}{6} = \frac{5}{6}.$$

The expansion of $(q + p)^4$ is

$$(q + p)^4 = q^4 + 4q^3p + 6q^2p^2 + 4qp^3 + p^4$$

Number of sixes obtained	Probability
0	$q^4 = \left(\dfrac{5}{6}\right)^4 = \dfrac{625}{1296}$
1	$4q^3p = 4 \times \left(\dfrac{5}{6}\right)^3 \times \dfrac{1}{6} = \dfrac{500}{1296}$
2	$6q^2p^2 = 6 \times \left(\dfrac{5}{6}\right)^2 \times \left(\dfrac{1}{6}\right)^2 = \dfrac{150}{1296}$
3	$4qp^3 = 4 \times \dfrac{5}{6} \times \left(\dfrac{1}{6}\right)^3 = \dfrac{20}{1296}$
4	$p^4 = \left(\dfrac{1}{6}\right)^4 = \dfrac{1}{1296}$

Note that the total probability covering all possible events is

$$\frac{625}{1296} + \frac{500}{1296} + \frac{150}{1296} + \frac{20}{1296} + \frac{1}{1296} = 1.$$

EXERCISE 19.

1. A card is cut from a pack of playing cards. Determine the probability that it will be a Jack or the Queen of Diamonds.

2. A coin is tossed and a die is rolled. Calculate the probabilities of:

   (a) a head and a six,   (b) a tail and an odd number.

3. A box contains 4 blue counters and 6 red ones. A counter is drawn from the box and then replaced. A second counter is then drawn. Find the probabilities that:

   (a) both counters will be blue,

   (b) both counters will be red,

   (c) one counter will be blue and the other red.

4. Out of 20 components 3 are defective. Two components are chosen at random from these 20 components. Determine the probability that they will both be defective.

5. A loaded die shows scores with these probabilities on the following page:

Score	1	2	3	4	5	6
Probability	0.15	0.17	0.18	0.13	0.16	0.21

(a) If I throw it once what is the probability of a score less than 3?

(b) If I throw it twice what is the probability of a 4 followed by a 6?

6. From a shuffled pack of cards, two cards are dealt. Find the probability of

(a) the first card being a King,

(b) the second card being a King if it is known that the first card was a King.

7. A B and C are points on a toy train system. The probability of going straight on at each point is $\frac{2}{3}$ (see Fig. 12.7). Find the probability that

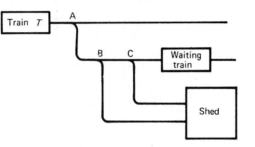

Fig. 12.7.

(a) the train T hits the waiting train,   (b) the train T goes into the shed.

8. *Event:* spin the pointer (Fig. 12.8)

(a) Find the probability that the pointer stops in B section.

(b) Find the probability that the pointer stops in R or G sections.

(c) The results of two successive spins are noted. Find:

    (i)   the probability that the first spin is in R and the second spin is in G.

    (ii)   the probability that the pointer stops in either R then G or G then R.

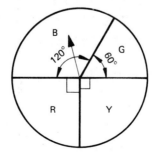

Fig. 12.8.

9. (a) If you cast an unbiased die what is the probability of it being a three?

(b) If you cast two such dice what is the probability of throwing (i) two threes (ii) a three and a four?

(c) Write down all the various ways in which a total of seven can be obtained with two dice. Use this information to calculate the probability of throwing a total of seven with two dice.

10. Two ordinary dice, one coloured red and the other blue, are thrown at the same time.

(a) What is the probability that the number on the red die will be 4?

(b) What is the probability that the number on the blue die will be even?

(c) Copy and complete the table below which shows the total scores on the two dice.

Number on red die

		1	2	3	4	5	6
	1	2	3	4	5	6	7
	2	3					8
Number on blue die	3	4					9
	4	5					10
	5	6					11
	6	7	8	9	10	11	12

(d) What is the probability of getting a total score of (i) 5, (ii) at least 9.

11. The letters of the word STATISTICS are written, one letter on each of ten cards. The cards are placed face downwards on a table, shuffled and one card is then turned face uppermost. What is the probability that it is

(a) a letter A,

(b) a letter S,

(c) not a vowel (i.e. not A or I).

The experiment is repeated but two cards are turned over together. What is the probability that

(d) both cards will be T's,

(e) one, at least, is a vowel.

12. A bag contains 7 red counters and 5 white counters. The counters are taken out in succession and not replaced.

(a) Copy and complete the tree diagram (Fig. 12.9) by writing in the correct fractions for boxes A, B, C and D.

(b) What is the probability of two red counters being taken out?

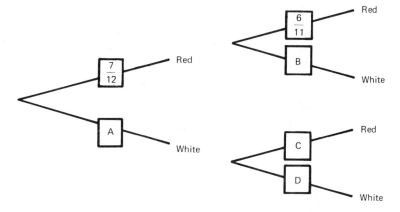

Fig. 12.9.

13. A box contains 3 red and 4 black balls. Draw a probability tree to show the probabilities of drawing one ball, then a second and finally a third, without replacement. From the tree answer the following questions:

(a) What is the probability of red, black, red?

(b) What are the chances of drawing red, red, black?

(c) What is the probability of drawing black, red, black?

(d) What is the probability of drawing only one black ball?

(e) What is the chance of drawing three black balls?

14. Fig. 12.10 shows a circle divided into five equal sectors which are numbered 1 – 5. A pointer attached to the centre of the circle is free to spin. A trial consists of spinning the pointer twice. The result of a trial will be (2,3) if the pointer stops at 2 on the first spin and on 3 on the second spin. The total for the trial is found by adding the two numbers together. Thus for the trial described the total is $2 + 3 = 5$.

(a) Show all the possible results of the trials.

(b) What is the probability of obtaining each of the following totals: 1, 2, 3, 6, 10?

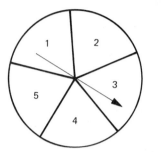

Fig. 12.10.

(c) How many possible different results of trials would there be if the circle was divided into:

    (i)   Two equal sectors numbered 1 and 2?

    (ii)   Three equal sectors numbered 1, 2 and 3?

    (iii) Four equal sectors numbered 1, 2, 3 and 4?

15.   Using Pascal's triangle obtain the expansion of the following:

    (a) $(q + p)^5$ (b) $(q + p)^7$ (c) $(q + p)^9$

16.   A coin is tossed four times. Calculate the probabilities of obtaining the following:

    (a) no heads (b) one head (c) two heads (d) three heads (e) four heads

17.   A pack of cards is cut twice. Calculate the probability that:

    (a) Two Aces will be obtained.

    (b) Two court cards will be obtained. (A court card is King, Queen or Jack.)

18.   5% of the fuses produced by a firm are known to be defective. A sample of four fuses is taken at random from a large batch of fuses. Calculate the probabilities that the sample will contain:

    (a) no defective fuses

    (b) two defective fuses

    (c) more than 1 defective fuses.

(Hint. Take $p = 5/100$ and $n = 4$ and use the method of repeated trials.)

13 ACQUIRING STATISTICS

13.1 INTRODUCTION

As we have already said, statistics is the name given to the science of collecting and analysing numerical facts. In this chapter we will discuss the methods used to collect statistical data.

There are two basic ways of acquiring statistical information:

(i) By using records already in existance.

(ii) By carrying out original research in the form of censuses, sample surveys, etc.

13.2 UNITS USED

Statistics involves measuring or counting. The figures obtained as a result of a survey must be expressed in units such as metres, litres, etc. The units used must be of a suitable size. Small objects, for instance, might be measured in centimetres but journeys made by a lorry would be measured in kilometres. Whenever possible standard units should be used. Units such as kilograms, seconds and metres, for instance, are suitable.

The units used should be of a constant size. As an example consider the production of a small factory. The figures might be: January – 43 000 units; February – 37 000 units; March – 44 000 units; April – 41 000 units. These figures are not directly comparable because the unit of time, the month, is not of a constant size. January and March have 31 days, April has 30 days and February has 28 days. A better unit of time might be 28 days.

13.3 QUANTITATIVE AND QUALITATIVE DATA

The data collected might be quantitative or qualitative. *Quantitative data* consist of facts in numerical form such as the heights of individuals, the number of driving licences issued, the diameters of ball bearings, etc. Qualitative data consist of facts like the colour of eyes, the taste of foods, the texture of cloth, etc. Qualitative data can often be put in quantitative form. Thus health can be measured by the number of days illness and intelligence by an IQ test. Qualities like taste and colour may be placed into order of rank. In the case of colour the lightest colour could be given rank 1, the next lightest rank 2, etc.

13.4 ACCURACY OF THE DATA

When the data are to be obtained by measuring, the type of equipment used can affect the accuracy of the measurement. For instance, a rule can measure to an accuracy of about 2 millimetres but a micrometer can be read to 0.01 mm. Optical equipment can be obtained which will measure accurately to 0.01 mm or less. Great acccuracy may therefore be obtained, but remember that extreme accuracy is very expensive to attain.

When counting, extreme accuracy may be misleading. For instance, during a Census of Population the population of a certain town was quoted as being 73 058 persons. This figure, even if it was correct at the time of publication was not likely to be maintained for any length of time because of births and deaths. We might say, therefore, that the population was about 73 000 or even, with some justification, about 70 000.

13.5 UP-TO-DATENESS OF THE DATA

The information used to make a decision should be up-to-date, but sometimes it is impossible to obtain figures which are up-to-date. For instance, the Census of Population takes between 6 months and 4 years to publish its findings.

13.6 USING RECORDS

Commercial organisations keep many records such as:

(i)	Financial	(v)	Labour turnover
(ii)	Sales	(vi)	Transport
(iii)	Production	(vii)	Stock control
(iv)	Wages and salaries	(viii)	Advertising costs

These records help in providing relevant, accurate and up-to-date information from which decisions and forecasts of future operations can be made.

13.7 PUBLISHED DATA

A great deal of statistical information can be obtained from published data which falls into two categories:

(i)   that published by the national and trade press

(ii)   that available in published reports issued by the Government.

13.8 NATIONAL AND TRADE PRESS PUBLICATIONS

The list below gives some details about a few of the national press publications.

(i) *The Financial Times* which is published daily. It contains statistical data in the form of charts, graphs and tables. It gives, daily, stock and share prices and it publishes a share price index known as The Financial Times Actuaries Share Index.

(ii) *Bank Reviews* published by Barclays, National Westminster, Midland and Lloyds. All of these contain statistical information of topical interest.

(iii) *International Review* which is published quarterly. It is published by Barclays Bank and it contains details of trade and economic conditions in many countries where the bank operates.

(iv) *The Economist* which is published weekly contains articles on business and economics. It also gives information on money and exchanges, and stock and share prices.

13.9 GOVERNMENT PUBLICATIONS

The most important sources of statistical information are the Government publications. The main ones are discussed below.

(i) *The Annual Abstract of Statistics* gives details of population, production, labour, prices and trade. In most cases figures for earlier years are included.

(ii) *The Monthly Digest of Statistics* is the monthly counterpart of the Annual Abstract of Statistics.

(iii) *Financial Statistics* is compiled by the Central Statistical Office (CSO) in collaboration with the Bank of England and other government departments.

(iv) *Economic Trends* is compiled by the CSO and published monthly. It contains statistics about production, manufacturing, building and construction, finance, trade, prices, wages, etc.

(v) *Statistical News* is compiled by the CSO and published quarterly. It contains accounts of current developments in official statistics and how they are compiled.

(vi) *Guide to Official Statistics* published occasionally by the CSO. The first issue was in 1976 and the second in 1978. It is a bibliography of statistical sources, official and non-official in the UK.

(vii) *Department of Employment Gazette* is compiled by the Department of Employment and published monthly. The contents include a summary of statistics relating to labour, unemployment, employment, unfilled vacancies, overtime, rates of pay, retail prices, etc. Special articles relating to the employment of labour are also included.

(viii) *Trade and Industry* compiled by the Department of Trade and published weekly. The main part of this publication gives very detailed figures on production, prices and trade.

(ix) *The Blue Book on National Income and Expenditure* compiled by the CSO and published annually. This book provides the main source of statistics about the Gross National Product, the Gross National Expenditure and the Gross National Income.

13.10 CONDUCTING A SURVEY

A survey may be carried out by either a *census* or a *sample* from the population. In a census the entire population under study is used but in sampling only a proportion of the population is used.

Sampling is used because sometimes it is not possible to survey the entire population. Problems arise with sampling because wrong conclusions about the population as a whole may be drawn by studying information obtained from only a fraction of the population. However, sampling has the following advantages over a census:

(i) it is cheaper because only a fraction of the population is surveyed

(ii) it may be carried out more quickly than a census

(iii) it can yield better information than can a census. This is because when sampling only a few interviewers are required whereas for a census many thousands may be used with the consequent difficulties of finding the right kind of personnel.

If a population is small a census is preferable. For instance, in a survey of working conditions in a factory a census is better because the number of people working in the factory (the population) is small.

13.11 THE SAMPLING FRAME

The sampling frame consists of all the items in the population. This frame is necessary so that any item in the population has a chance of becoming part of the sample.

13.12 METHODS OF COLLECTING DATA

Information can be collected by several different methods as follows:

(i) *By direct observation* (e.g. counting the number of vehicles passing a certain point in a given time). This method reduces the chance that incorrect information may be garnered, but it is not always feasible (e.g. it would be practically impossible to follow a housewife for a month to find out what she bought when out shopping).

(ii) *By personal surveys.* The most common way of obtaining information in such fields as market research is the face-to-face interview. The trained interviewer, using a questionnaire, asks questions of individuals and notes the answers. This method has the advantage that many questions can be asked quickly and that high response rates are achieved. However, an interviewer can influence the answers and this may introduce a systematic bias into the survey.

(iii) *Postal surveys* use a sample of people drawn from a specific mailing list or from an electoral register. The people selected are sent a questionnaire. This method has the following advantages:

(a) interviewing bias is avoided

(b) the respondents can take their time answering and thus give more consideration to the answers

(c) costs are generally low.

However, postal surveys have several disadvantages as follows:

(a) low response rates which may cause bias

(b) the length of time needed for the survey

(c) lack of questionnaire control. Different people might interpret a question in different ways, something that does not happen when an interviewer is used.

(iv) *Telephone surveys* which are special cases of personal interviews. These are becoming more widely used in the UK because more and more people have a telephone at home.

13.13 THE PILOT SURVEY

This is essentially a small scale replica of the actual survey and it is carried out before the actual survey is undertaken. It should duplicate, as near as possible, the survey which is to be made because it may reveal snags in the proposed questions and methods. A pilot survey is very useful when the actual survey is to be on a big scale as it may provide data which will allow costs to be trimmed. Also, a pilot survey will give an estimate of the non-response rate and it will also give a guide as to the adequacy of the sampling frame chosen.

13.14 DESIGNING THE QUESTIONNAIRE

There are no hard and fast rules which have to be obeyed when designing a questionnaire. Nevertheless the following points should be noted:

(i) The questions should be simple and unambiguous. Long questions containing more than one element should be broken down into sub-topics.

(ii) Leading questions such as 'Don't you agree that all intelligent people read the XYZ magazine' should be avoided because the answer is suggested. To the question posed, the answer 'yes' is strongly suggested.

(iii)   Embarrassing or irrelevant questions should be avoided.

(iv)   The questions should be as short as possible and they should be asked in a logical sequence because the quality of the answers will then be improved.

(v)   Questions which allow an answer to be ticked are best.

13.15 SAMPLING METHODS

Random sampling

1. RANDOM SAMPLING ensures that each member of a population has an equal chance of being chosen. This is the only kind of sample in which the selection is free from bias. However, a random sample will not, necessarily, give the best cross-section of a population and it does not guarantee freedom from bias – it is only the selection which is bias-free.

Non-random sampling

2. NON-RANDOM SAMPLING is used when it is not feasible to use a random sample. In such cases one of the following methods is used:

Multi-stage sampling in which the country is divided up into a number of geographical areas. Three or four of these are selected at random. Each selected area is then split up into smaller areas and a few of these small areas are then randomly selected. From each small area the individuals to be quizzed are selected at random. This method is used when a random sample would entail a great deal of travelling on the part of interviewers.

Quota sampling in which the interviewer is not given the names of the individuals to be seen but instead is told to interview a number of people at his own discretion. The quota is nearly always divided up into different types of people (e.g. managers, teachers, skilled workers, labourers, etc.).

Cluster sampling. This method is used when the size of the population is unknown as would be the case with trees and other plants, animals and fish. The enumerators go to small areas, chosen in the same way as for multi-stage sampling, and count every item fitting the given description.

Systematic and stratifed sampling

3. SYSTEMATIC SAMPLING involves taking a percentage of the total population. If, for instance, a 10% sample is required then every tenth item of the listed population is taken. The method gives virtually a random sample and provided there is no periodic pattern, bias is eliminated. However, in some industrial applications such as quality control, the method should not be used. For example in a machine having, say, six heads if we take every sixth item produced we shall, in effect, be sampling only from one of the heads. What is happening to the other heads will never be revealed.

4. STRATIFIED SAMPLING. A population may be divided up into sub-groups whose members have more in common with one another than they have with the population as a whole. Thus people fall into sub-groups according to sex, age, social background, etc. A survey on pensions may well draw different opinions and answers from different sub-groups. Suppose in a certain city 32% of the population was under 20 years of age, 26% between 20 and 40 years old, 24% between 40 and 60 and 18% over 60 years old. To stratify the sample by age in a sample of 1000 individuals we should interview

> 320 people under 20 years old
>
> 260 people between 20 and 40 years old
>
> 240 people between 40 and 60 years old
>
> 180 people over 60 years old.

The individuals in each strata are randomly selected.

EXERCISE 20

1. Explain what is meant by sampling in statistics. Describe the following types of sample:

   (a) systematic   (b) stratified   (c) multi-stage   (d) cluster.

   In what kinds of enquiry are the above forms of sampling appropriate (one example of each type of sample is sufficient).

2. Give brief explanations of the following terms used in sample surveys:

   (a) interviewer bias   (b) non-response   (c) sampling frame   (d) pilot survey.

3. Discuss the relative advantages and disadvantages of the personal interview and the postal questionnaire as a method of collecting data.

4. Describe in some detail two of the following:

   (a) General Index of Retail Prices

   (b) Index of Industrial Production

   (c) Unemployment Statistics

   (d) Statistics on Wages and Salaries

5. Compare and contrast different methods of gathering information in survey methodology.

6. Questionnaire design is one of the main steps undertaken in the construction of a survey. Outline the characteristics of a good questionnaire and explain the role of the pilot survey in helping questionnaire design.

7. What are the principal sources of statistical information on incomes in the UK?

8. (a) Explain carefully what is meant by stratified sampling and why it is often preferred to either random or purposive sampling.

   (b) Explain briefly why non-response is such a problem in postal surveys.

   (c) Some of the children at a school in a large town are engaged in a project which, in part, is concerned with the newspapers purchased and read by the town's population. In order to obtain a sample representative of the town they decide to ask every child in the school to report which newspapers are regularly purchased by his or her family. Assuming total response comment on the possible existence of bias.

14 TABULATION

14.1 RULES FOR TABULATION

Information given in narrative form is often difficult to understand. Consider the following statement:

'In 1951, 207 thousand persons received unemployment benefit, 906 thousand persons sickness benefit, 1437 thousand males retirement pensions, 2709 thousand females retirement pensions, 457 thousand received widows benefit. 217 thousand persons received other National Insurance benefits. In 1971 the corresponding figures for unemployment benefit was 457 thousands, for sickness benefit was 969 thousands, for male retirement pensions 2611 thousands, for female retirement pensions 5196 thousands, for widows benefit 448 thousands, for other National Insurance benefits 387 thousands.'

The narrative is difficult to read and very confusing. If the information is tabulated as below the confusion is eliminated.

National Insurance Benefits

Type of benefit	Numbers obtaining benefits (thousands)	
	In 1951	In 1971
Unemployment	207	457
Sickness	906	969
Retirement pensions (males)	1437	2611
Retirement pensions (females)	2709	5196
Widows pensions	457	448
Others	217	387

In this table we have only given the figures stated in the narrative. It may help the reader who is interested in the data if we add some secondary statistics such as the total number of people drawing benefits and percentages of the totals as shown on the following page.

National Insurance Benefits

Type of benefit	1951		1971	
	Numbers drawing benefit (thousands)	% of total	Numbers drawing benefit (thousands)	% of total
Unemployment	207	3.49	457	4.54
Sickness	906	15.27	969	9.62
Retirement pensions (males)	1437	24.22	2611	25.93
Retirement pensions (females)	2709	45.66	5196	51.61
Widows pensions	457	7.70	448	4.45
Others	217	3.66	387	3.85
Totals	5933	100	10068	100

14.2 PRINCIPLES OF TABLE CONSTRUCTION

When attempting to construct a table the following principles should be observed:

(i) The original figures should be presented in an orderly fashion so that any pattern emerging from them can be easily seen.

(ii) Where it is thought necessary the original figures can be summarised.

(iii) Try to keep the tabulation as simple as possible.

(iv) The table with its headings should be self-explanatory.

(v) The source of the figures should be stated if known.

(vi) Always state the units used.

(vii) Where appropriate, totals, sub-totals and percentages can be shown.

(viii) Figures likely to be compared should be placed near one another.

(ix) To eliminate unnecessary detail rounding should be used.

EXAMPLE 1. In England in a certain year there were 16 070 000 dwellings: of these 8 360 000 were owner occupied, 4 500 000 were rented from local authorities, 2 410 000 were rented from private owners, the remainder being held under other tenures. In Scotland in the same year there were 1 800 000 dwellings: of these 540 000 were owner occupied, 940 000 were rented from local authorities, 200 000 were rented from private owners, the remainder were held under other tenures.

Tabulate these data, calculate appropriate secondary statistics and include these secondary statistics in your tabulation.

Housing in England and Scotland

Type of house	England		Scotland	
	Number (thousands)	Percentage of total	Number (thousands)	Percentage of total
Owner occupied	8360	52	540	30
Rented from local authorities	4500	28	940	52
Rented from private owners	2410	15	200	11
Held under other tenures	800*	5*	120*	7*
Totals	16070	100	1800	100

The figures marked with a star and the corresponding percentages were not given in the narrative and these have been calculated.

Sometimes the figures given in a narrative can be extremely bewildering and when tabulating several arithmetic calculations may be necessary as shown in Example 2.

EXAMPLE 2. 4629 students sat examinations in single-subject courses comprising Statistics, Mathematics and Computer Studies. The examinations were at O-level, A-level and S-level. 2216 students out of the 4629 took O-level subjects and 1193 A-level. 1943 took the examinations in Statistics and 1147 the examinations in Mathematics. 984 students took the O-level in Statistics and 765 the S-level in that subject. 334 took A-level Mathematics and 665 A-level Computer Studies. 195 students sat the S-level examinations in Mathematics.

Draw up a suitable table to show the above data.

Let us start by tabulating the figures as they stand as shown overleaf.

Numbers of Candidates taking examinations in
Statistics, Mathematics and Computer Studies

Subject	O-level	A-level	S-level	Totals
Statistics	984		765	1943
Mathematics		334	195	1147
Computer Studies		665		1539
Totals	2216	1193		4629

As can be seen, there are several gaps in the table. We can fill these by addition and subtraction and so complete the table as shown below:

Number of Candidates taking examinations in
Statistics, Mathematics and Computer Studies

Subject	O-level	A-level	S-level	Totals
Statistics	984	194	765	1943
Mathematics	618	334	195	1147
Computer Studies	614	665	260	1539
Totals	2216	1193	1220	4629

14.3 FALSE CONCLUSIONS

At the end of a statistical report it is usual to draw conclusions. Unless care is taken the wrong conclusions may be drawn from a set of figures. Consider the statement: 'The crude death rate in Bournemouth is much higher than in some inland cities. Therefore it is untrue to say that seaside resorts are good for health.'

On the face of it the conclusion seems plausible but it goes against the fact that seaside places are usually healthy places to live in because there is less air pollution, etc. Why should the crude death rate be higher in Bournemouth? Is it anything to do with the type of population living in Bournemouth? We now realise that many elderly people go to Bournemouth to retire and because the death rate for old people is higher than that for the population as a whole the crude death rate for Bournemouth would be expected to be high. It is unfair in this case to relate health and the crude death rate. The standardised death rate would probably reveal a different state of affairs.

EXERCISE 21

1. At the end of 1973 the total value of UK external assets was £65 075 million whilst the total of external liabilities was £59 600 million. The assets included private investment abroad of £19 725 million; UK banks claims abroad of £39 384 million; UK Government loads of £1 790 million and official reserves of £2 237 million. The liabilities included private investment in Britain £10 405 million, public sector borrowing £5 285 million, UK banks liabilities to overseas customers £42 288 million.

(Source: Bank of England)

Tabulate these data and calculate suitable secondary statistics which should be incorporated into the tabulation.

2. In the UK in 1973, 1 192 000 people were found guilty of motoring offences compared with 291 000 in 1951. 97 000 were found guilty of drunkenness compared with 51 000 in 1951. 88 000 were found guilty of offences concerned with motor vehicle licences compared with 6 000 in 1951. 43 000 were found guilty of offences concerned with Wireless and Telegraphy Acts compared with 3 000 in 1951. 159 000 were found guilty of other offences compared with 213 000 in 1951.

Tabulate these data and calculate any appropriate secondary statistics.

3. *What Children eat*

The fifth form pupils in a Midlands School carried out a survey into the eating habits of the 600 pupils in the school.

This survey showed that only one-third of the school ate a cooked breakfast, most eating toast and/or cereal. 47.5% of the boys and 41.2% of the girls had school dinners. Half of those who went home for lunch had a cooked meal (meat and 2 veg), the rest having a cooked snack, for example beans on toast.

In the evening 48.5% of boys and 51% of girls had a 'dinner' meal (cooked meat + 2 veg type). 13.5% of boys and 15.7% of the girls had only a tea meal, i.e. bread and butter, jam and cakes, while 38% of boys and 33.3% of girls had a 'high' tea.

60.6% of boys and 70.6% of girls had supper, but two-thirds of these had only biscuits, crisps or sweets. Only 8% of boys and 12% of girls had a milk drink at night.

Tabulate these figures calculating any extra entries where necessary. (Assume that the total number of pupils is evenly split between boys and girls.)

4. The number of unemployed, excluding school leavers and adult students seeking employment in Great Briatin on 9th September of a certain year, was 584 384; this represents an increase of 11 697 over the August figure. In addition, there were 33 426 unemployed school leavers and 29 301 unemployed adult students, so that the total number of unemployed was 647 111, a fall of 9 186 since August. During August, the number of unemployed school leavers was 55 976 and there were 27 634 unemployed adult students.

(Source: LCC)

Present this information in the form of a statistical table, and calculate suitable secondary statistics to illustrate the nature of the information.

5. (a) What are the basic rules to be observed when designing a table for the presentation of statistical data?

   (b) Design a blank table to show a breakdown by ages of the estimated population (at 30 June 1978) of the UK. Show totals of persons, males and females of the UK, males and females of England and Wales, males and females of Wales only, males and females of Scotland and males and females of Northern Ireland.

6. Criticise the following statements as evidence of statistical facts and explain fully how each one could be misleading as to their conclusions.

   (a) Traffic fatalities increased from 15 to 45, an increase of 300 per cent.

   (b) The average age at which girls in Hopeville begin to use lipstick is 11.897 years.

   (c) A class in Psychology, after listening to a recording of Brahms' Symphony No. 4 in E Minor and Beethoven's Symphony No. 3 in E Flat Major, indicated the following preferences: Brahms, 32; Beethoven, 12. This shows clearly that people prefer Brahms to Beethoven.

   (d) A company decides to find out whether radio advertising increases sales. After an intensive radio advertising programme of one month it is observed that sales have increased over the level of the previous month. It is concluded that the increase was due to the radio advertising.

7. Comment critically on the following statements showing (with reasons) whether you would accept them or consider them to be wrong.

   (a) The illegitimacy rate in London is the highest in the country, which proves that large cities are immoral places.

   (b) More people were killed in air accidents in 1978 then in 1928. Therefore it was more dangerous to fly in 1978 than in 1928.

   (c) There is a greater incidence of rheumatism in Bath per head of the population than in Birmingham. Therefore the Birmingham environment is less likely to cause rheumatism than that of Bath.

MISCELLANEOUS EXERCISE

EXERCISE 22

(All questions are of the type found in O-level, CSE and RSA examination papers.)

Short answer questions

1. An article cost £128 in December, 1977. By December 1978 its price had increased by £16. From December 1978 to December 1979 the price increased by 20%. Calculate the December 1979 index number (1977 = 100).

2. The table below gives corresponding values of the two variables x and y.

x	2	4	6	10	15
y	7	13	19	31	46

    Determine the coefficient of correlation between x and y.

3. The table below shows the price of a certain article in each of the years 1976 – 79.

Year	1976	1977	1978	1979
Price	£4.00	£4.80	£6.00	£6.60

    Using 1976 as base calculate the index numbers for 1977, 1978 and 1979 using the chain base method.

4. The table below shows the marks awarded to two ice skaters by two judges. Calculate the value of the coefficient of rank correlation.

Skater	A	B	C	D	E	F
1st judge	5.5	6.0	6.0	7.0	7.5	8.0
2nd judge	6.5	8.0	7.0	5.5	7.5	7.5

5. Each of five essays was given a mark out of 10 by two independent examiners.

Essay	A	B	C	D	E
1st examiner	9	7	10	4	5
2nd examiner	10	6	8	5	9

Calculate the value of the coefficient of rank correlation.

6.

Commodity	A	B	C	D
Index	106		114	100
Weight	2	3	3	1

If the weighted index for the four commodities is 111, determine the index for commodity B.

7. Show, on a diagram, points which will indicate that perfect negative correlation exists between two variables x and y.

8. The index number of a piece of equipment was 140 in 1977, taking 1975 as the base year. In 1977 its price was £9800. What was its price in 1975?

9. Two variables have positive linear correlation. Write down the values between which the coefficient of correlation must lie.

10. The corresponding values of two variables x and y are plotted on a graph. It is found that the regression line of y on x has a gradient of -3 and that $(\bar{x}, \bar{y})$ is $(2,9)$. Estimate the value of y when $x = 3.5$.

11. Find the third average in a three-point moving average for the values:

   14, 23, 18, 12, 15, 24, 30, 19, 15

12. The values of the variables x and y are such that $\bar{x} = 8$ and $\bar{y} = 5$. The regression line of y on x passes through the point $(6,1)$. Find the equation of the regression line of y on x.

13. From a set of 20 observations it is found that the coefficient of regression of y on x is -2.6. By using the regression line of y on x it is found that when $x = 6$, $y = 9.9$. Given that the sum of the x values is 60, what is the sum of the y values?

14. Why does non-response cause difficulty in a sample survey?

15. What is the value of the second average in a four-point moving average for the values

    25, 23, 29, 22, 27, 22, 28, 30, 26 ?

16. Fig. 1 shows the frequency distribution of goals scored by the school hockey team. With the aid of this chart find

    (a) the total number of goals scored

    (b) the mean number of goals per match

    (c) the probability that a spectator who watched only one match, selected at random, saw at least two goals scored by the team.

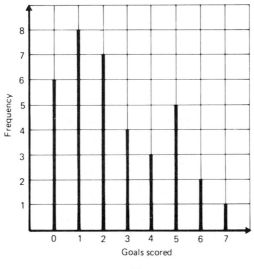

Fig. 1.

Standard questions

17. Corresponding values of two variables, x and y, are shown in the table below:

x	13.0	20.7	29.0	17.0	25.0	21.5
y	22.0	34.6	46.0	28.0	40.0	33.6

    (a) Calculate Spearman's coefficient for the rank correlation between x and y.

    (b) Plot a scatter diagram of the data.

(c) Calculate the arithmetic mean values of x and y and use them to draw the line of best fit.

(d) Determine the equation of this line in the form $y = ax + b$ stating the values of a and b correct to two significant figures.

(e) Use this equation to estimate the value of y when $x = 19.0$.

18. The table below shows the average weekly sales of two firms X and Y during the period 1977 – 79:

	1977			1978		
	Jan – Apl	May – Aug	Sep – Dec	Jan – Apl	May – Aug	Sep – Dec
X	2650	2550	3050	5050	4800	5300
Y	5050	2700	4150	5800	4450	4960

	1979		
	Jan – Apl	May – Aug	Sep – Dec
X	7300	7050	7550
Y	6550	5200	5650

(a) Using 2 cm to represent £1000 and 2 cm to represent 4 months, plot both sets of information on the same axes.

(b) For both firms, calculate an appropriate moving average so as to eliminate seasonal variation from the sales.

(c) Plot the two moving averages on the graph.

(d) Estimate the average weekly sales of firm X during the first four months of 1980.

19. The marks obtained by each of ten candidates in two tests are given below.

Maths	64	54	66	72	50	78	74	48	41	68
French	47	54	47	46	55	40	38	58	65	43

(a) Use these figures to plot the points on a scatter diagram, using scales of 2 cm to represent 10 marks on both axes, and draw in a line of best fit.

(b) State what conclusion you can draw about the correlation between the two sets of marks.

(c) Another candidate obtained 60 marks in Maths but was absent from the French examination. Estimate a French mark for this candidate.

20. *Sales of Gramophone records*

Millions				
Quarter	1	2	3	4
1976			24	33
1977	28	27	25	33
1978	30	28	27	36
1979	30	29		

(a) By means of a moving average find the trend.

(b) Plot the original data and the trend on the same graph paper.

21. Consider the set of numbers 5, 7, 10, 11, 16, 19.

(a) Giving the answers as fractions in their lowest terms, find the probability that a number chosen at random from the given set is

(i) even

(ii) odd

(iii) prime

(iv) an exact square.

(b) Find the probability that two numbers chosen separately at random from the set without replacement will each be greater than 10.

22. A bag contains counters: 14 yellow counters and 10 red counters. Two counters are taken out in succession and not replaced.

(a) Copy and complete the tree diagram (Fig. 2) by writing in the answer spaces opposite, the correct fractions for boxes A, B C and D.

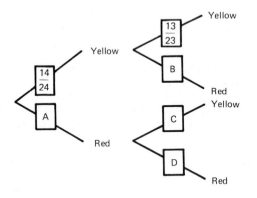

Fig. 2.

(b) What is the probability that two red counters are taken out?

23. The number of children present in a class of 31 each day over a period of 3 weeks was as follows:

Week	M	T	W	T	F
1	30	31	29	28	27
2	31	31	29	29	26
3	30	29	30	28	25

(a) Illustrate the above information by means of a suitable line graph.

(b) Calculate and tabulate, the '5-day moving averages'.

(c) Superimpose on your first graph, a graph of the '5-day moving averages'.

(d) Comment on the results.

24. (a) *Jallop Poll*

   Question: Do you agree with requisitioning empty houses for the homeless?

   Sample: 50 people on a 'Shelter' walk.

   Result: 45 people said 'Yes'.

   (i)   Was this a truly random sample?

   (ii) What percentage of the sample said 'No'?

   (iii) A newspaper headline said:

   '90% of the country support requisitioning empty houses for the homeless'.

   Give *two* reasons why this is not a fair conclusion from the sample.

(b) *Advertisement*

   'Public prefers ALPHA ice cream machine'.

   '1,000 people voted'.

   Facts: 1000 people were asked:

   Do you use an ice cream machine?

   Answer: YES (100)   NO (900)

   Which one do you use?

   Answer: ALPHA (30)   A (25)   C (25)   D (20)

   (i) What size was the sample of users of ice cream machines?

   (ii) Is it a fair claim from these facts that 'Public prefers an ALPHA ice cream machine'? Give reasons for your answer.

25. The following table gives the number of holiday bookings (in hundreds) taken by a travel agent over a period of four years.

| | Quarter | | | |
	1st	2nd	3rd	4th
1975	28	26	23	23
1976	25	25	21	18
1977	21	21	18	12
1978	16	16	15	9

(a) Plot these values on graph paper and join successive points with straight lines.

(b) Calculate the yearly (that is 4-quarterly) moving averages.

(c) Plot these 'average' points carefully on the same graph.

(d) Comment briefly on the general trend implied by the moving average graph of (c).

(e) What is the purpose of plotting such a graph as (c)?

26. (a) Which of the following do *not* represent a coefficient of correlation: 1.6, 0.8, 0.02, -0.3, -0.99, -9.9?

(b) Ten shades of the colour green when arranged in their true order from light to dark are numbered:

1, 2, 3, 4, 5, 6, 7, 8, 9, 10 respectively.

An observer when asked to arrange the shades from light to dark (after they had been mixed up) produces the following rank:

3 1, 5, 2, 6, 4, 10, 9, 8, 7.

Calculate a coefficient of rank correlation for these data (correct to two decimal places), and comment on the result.

27. Two dice are thrown. Each die has its faces numbered 1 to 6.

(a) Copy and complete the table which shows the possible scores obtained when the two numbers on the uppermost faces are added.

Score on Die A

		1	2	3	4	5	6
	1						7
	2	3		5			
Score on Die B	3					8	
	4		6				
	5				9		
	6						

(b) What is the probability of obtaining a total score of 3 with the two dice?

(c) What is the probability of obtaining a total score greater than 9?

(d) What is the probability that neither die shows a one on the uppermost face?

(e) A game with two dice is played at a side show of a garden fete. It costs 1 p to roll the two dice, but you receive 3 p if you score 10, 4 p if you score 11 and 6 p if you score 12. If, over the course of the fete, 720 people each had a turn, what would you expect the profit to be on the stall?

28. A reading test was given to ten groups of children, there being 100 children in each group. Each child read out a list of words and a record was kept of the number he read correctly. The average age of the first group was 8 years, that of the second group 8½ years, that of the third group 9 years, and so on. Each group of children was claimed to be a representative cross-section of children of a particular age group. The table below shows the *total* number of words read correctly by the 100 children in each group.

Group	1st	2nd	3rd	4th	5th
Average age (in years)	8	8½	9	9½	10
Total number of words correctly read by 100 children	2928	3343	3682	3968	4229

Group	6th	7th	8th	9th	10th
Average age (in years)	10½	11	11½	12	12½
Total number of words correctly read by 100 children	4452	4886	5119	5433	5704

Plot a scatter diagram and draw a line of best fit. Use your line of best fit to estimate the probable number of words correctly read by a child of average reading ability and of age:

(a) 8 years 4 months,

(b) 10 years 8 months.

Clearly indicate on your diagram any necessary reading made to help you obtain each answer and comment on any assumptions you make.

29. The table below shows the number of deaths occurring in a particular year in each of two towns, X and Y, together with the population of each town and of the country; all are classified by age group.

Age group (years)	Number of deaths		Population		
	Town X	Town Y	Town X	Town Y	Country (millions)
0 –	24	48	3000	8000	15
20 –	9	20	3500	7500	14
40 –	36	77	2500	3500	13
60 and over	90	82	1000	1000	8

Calculate crude and standardised death rates for each town.

30. The following table shows experimental values of two variables x and y.

x	3	5	8	12	17
y	7.3	8.7	10.4	12.5	14.3

It is expected that the relationship between y and x is of the form $y = a\sqrt{x} + b$ where a and b are constants.

Plot y against $\sqrt{x}$ and draw the line of best fit. Obtain the equation of this line. Use the relationship to predict a value of y when $x = 81$.

31. (a) A box contains 4 red balls, 3 green balls and 2 blue balls. A ball is taken from the box. What is the probability that the ball removed is

   (i) red?

   (ii) green?

   (iii) blue?

   (b) The ball is now replaced and a ball is again taken from the box.

   (i) Draw a tree diagram to show the possible outcomes.

   (ii) Hence use your diagram, or otherwise, to find the probability of taking out:

   (1) 2 balls of the same colour in succession.

   (2) 2 balls of different colours in succession.

(c) Another box contains some white, some yellow and some black balls. If a ball is drawn at random from the box the probability it is white is ⅓ and the probability it is black is ¼ . If there are 15 yellow balls in the box, how many white ones are there?

32. *Cost of food in Ruritania*

The main items of diet are Rusks, Rissoles and Reba. The currency is Rins.

Table A	Base year 1973	1974	% change	Weights	Weight X %
Rusks (packet)	8 rins	10 rins		A =	
Rissoles (pack)	20 rins	24 rins	20%	B =	
Reba (bottle)	10 rins	11 rins		C =	

(a) Copy and fill in the remaining % changes in Table A.

(b) The weights are based on a weekly expenditure in 1973 on:

Table B	Cost in Rins	Suitable weights
10 packets of Rusks		A =
3 packs of Rissoles		B =
4 bottles of Reba		C =

Work out the cost of the three items and from these choose suitable weights A, B and C.

(c) Fill these in Table A and use them to calculate the cost of food index for 1974 compared with 1973.

33. The following table gives the marks of ten students in each of two examinations.

Maths	70	39	61	49	64	34	42	72	52	57
Physics	73	44	62	54	70	41	46	76	60	64

(a) On graph paper plot a scatter diagram.
(Plot Maths horizontally and Physics vertically.)

(b) Calculate the mean mark for each examination and mark this point on the graph.

(c) Draw, on the graph, a line for best fit.

(d) Use your graph to estimate the mark of a student who missed the Maths examination but had 50 marks in the Physics examination.

MULTIPLE CHOICE QUESTIONS

34. In 1974 an article cost £42. The price relative in 1979 (1974 = 100) is 125. The cost of the article in 1979 is

a £33.60   b £52.50   c £67   d £62.50

35.

Month	Jan	Feb	March	April	May
Sales	120	130	110	180	150

For the above data the second three-monthly moving average is

a 120   b 140   c 230   d 138

36. r is the correlation coefficient. One of the following statements is false. Which one?

a $r = 0.8$ b r is less than 0.7   c r is greater than 1   d $r = 1$

37. A card is chosen at random from a pack of 52 playing cards. The probability of it being a Queen is

a $\dfrac{1}{4}$ b $\dfrac{1}{8}$ c $\dfrac{1}{13}$ d $\dfrac{1}{52}$

38. There are two black balls and one white ball. Two of the three are chosen at random. The probability that both will be black is

a $\dfrac{2}{3}$ b $\dfrac{1}{6}$ c $\dfrac{1}{3}$ d $\dfrac{2}{9}$

39. An index number for 1979 (1976 = 100) is 125. The index number for 1976 with 1979 as base is

a 80   b 75   c 225   d 62.5

40. The diagram (Fig. 3) shows a scatter diagram. The correlation coefficient between x and y could be

    a 1   b − 1   c − 0.8   d 0.8

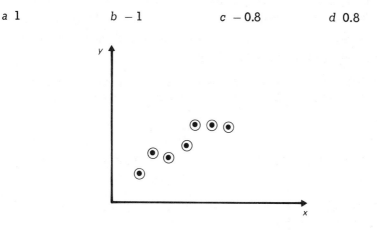

Fig. 3.

41. Eight students took examination papers in Statistics and Physics. The students marks for each paper were ranked. If d is the difference between ranks and $\Sigma d^2 = 48$, then the coefficient of rank correlation is

    a − 0.12   b − 0.10   c 0.87   d 0.43

42. If m is the probability of an event and n is the probability of another independent event, then the probability of both events happening is

    a $m + n$ b $1 - (m + n)$ c $1 - mn$ d mn

43. The following figures are the probabilities of certain events happening. There must be an error in one of the probabilities. Which?

    a 0   b 1   c − 0.5   d 0.6

44. The table below gives corresponding values of two variables x and y. It is essential that the line of best fit passes through one of the points given. Which one?

x	2	3	6	7	10	14
y	6	9	16	17	24	30

    a (2, 6)   b (12, 26)   c (7, 17)   d (6, 16)

45. A firm exported 5000 units in 1976 and 8000 units in 1979. The quantity relative for 1976 (1979 = 100) is

    a 80   b 50   c 160   d 62.5

46. If p is a probability one of the following is necessarily true. Which?

a p is negative   b p is positive   c p is greater than 1
d p lies between 0.5 and 1

47. The diagram (Fig. 4) shows a scatter diagram with the line of best fit drawn. The equation for the line of best fit is

a $y = 2x - 3$ b $y = 3 - 2x$ c $y = 2x + 3$ d $y = 15x + 6$

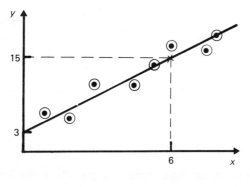

Fig. 4.

48. One of the following is not a component of a time series. Which?

a Trend   b cyclical variation   c long-term variation
d irregular variation

49. In a time series the secular variation is

a seasonal variation

b cyclical variation

c long term variation

d irregular variation

50. The table below shows how prices have varied over a period of years.

Year	1975	1976	1977	1978	1979
Price	20	24	30	36	72

Using a chain base index (1975 = 100) the index number for 1978 is

a 120   b 150   c 180   d 200

MATHEMATICAL APPENDIX

SUBSCRIPT NOTATION

Let x_i denote any of the numbers $x_1, x_2, x_3, \ldots, x_n$ assumed by a variable x. The letter i in x_i is called the subscript of x and it can stand for any of the numbers $1, 2, 3, \ldots, n$.

SUMMATION NOTATION

The symbol $\sum\limits_{i=1}^{n} x_i$ is used to denote the sum of all the x's from $i = 1$ to $i = n$. That is,

$$\sum_{i=1}^{n} x = x_1 + x_2 + x_3 + \ldots + x_n$$

Where there is no danger of confusion we will denote this sum simply by Σx. Thus

$$\Sigma x = x_1 + x_2 + x_3 + \ldots + x_n$$

$$\sum_{i=1}^{n} ax_i = ax_1 + ax_2 + ax_3 + \ldots + ax_n = a \, \Sigma x$$

$$\sum_{i=1}^{n} a = a + a + a + \ldots + a = na$$

$$\sum_{i=1}^{n} f_i x_i = f_1 x_1 + f_2 x_2 + f_3 x_3 + \ldots + f_n x_n$$

$$\sum_{i=1}^{n} (a + x_i) = (a + x_1) + (a + x_2) + (a + x_3) + \ldots + (a + x_4) = na + \Sigma x$$

$$\sum_{i=1}^{n} (x_i - a)^2 = \sum_{i=1}^{n} (x_i^2 - 2ax_i + a^2)$$

$$= (x_1^2 - 2ax_1 + a^2) + (x_2^2 - 2ax_2 + a^2) + \ldots + (x_n^2 - 2ax_n + a^2)$$

$$= \Sigma x^2 - 2a \, \Sigma x + na^2$$

1. To show that $\bar{x} = A + \dfrac{\sum\limits_{i=1}^{n} d_i}{n}$

Let the numbers $x_1, x_2, x_3, \ldots, x_n$ have deviations $d_1, d_2, d_3, \ldots, d_n$ from any number A. That is,

$$d_1 = x_1 - A, \quad d_2 = x_2 - A, \quad d_3 = x_3 - A, \ldots, d_n = x_n - A$$

Thus $x_i = A + d_i$

$$\bar{x} = \frac{\sum\limits_{i=1}^{n} x_i}{n} = \frac{\sum\limits_{i=1}^{n} (A + d_i)}{n} = \frac{nA + \sum\limits_{i=1}^{n} d_i}{n}$$

$$= A + \frac{\sum\limits_{i=1}^{n} d_i}{n}$$

2. To show that $\bar{x} = A + \dfrac{\sum\limits_{i=1}^{n} f_i d_i}{n}$

As before, $x_i = A + d_i$

$$\bar{x} = \frac{\sum\limits_{i=1}^{n} f_i x_i}{n} = \frac{\sum\limits_{i=1}^{n} f_i (A + d_i)}{n}$$

$$= \frac{\sum\limits_{i=1}^{n} f_i A + \sum\limits_{i=1}^{n} f_i d_i}{n}$$

$$= \frac{A \sum\limits_{i=1}^{n} f_i}{n} + \frac{\sum\limits_{i=1}^{n} f_i d_i}{n}$$

$$= \frac{An}{n} + \frac{\sum\limits_{i=1}^{n} f_i d_i}{n}$$

$$= A + \frac{\sum\limits_{i=1}^{n} f_i d_i}{n}$$

Since $\sum\limits_{i=1}^{n} f_i = f_1 + f_2 + f_3 + \ldots + f_n = n$

3. To show that $\sigma = \sqrt{\dfrac{\Sigma fd^2}{n} - \overline{d}^2}$

As before, $x_i = A + d_i$ and $\overline{d} = \dfrac{\Sigma fd}{n}$

$$\sigma = \sqrt{\frac{\Sigma fx^2}{n} - \overline{x}^2}$$

$$= \sqrt{\frac{\Sigma f(A + d)^2}{n} - \left(\frac{\Sigma f(A + d)}{n}\right)^2}$$

$$= \sqrt{\frac{\Sigma f(A^2 + 2Ad + d^2)}{n} - \left(\frac{\Sigma fA + \Sigma fd}{n}\right)^2}$$

$$= \sqrt{\frac{A^2 \Sigma f}{n} + \frac{2A \Sigma fd}{n} + \frac{\Sigma fd^2}{n} - \left(\frac{nA}{n} + \frac{\Sigma fd}{n}\right)^2}$$

$$= \sqrt{A^2 + 2A\overline{d} + \frac{\Sigma fd^2}{n} - (A + \overline{d})^2}$$

$$= \sqrt{A^2 + 2A\overline{d} + \frac{\Sigma fd^2}{n} - A^2 - 2A\overline{d} - \overline{d}^2}$$

$$= \sqrt{\frac{\Sigma fd^2}{n} - \overline{d}^2}$$

4. If the numbers $x_1, x_2, x_3, \ldots, x_n$ are all increased by the same amount, a, such that $X_1 = a + x_1$, $X_2 = a + x_2$, $X_3 = a + x_3, \ldots,$ $X_n = a + x_n$, then

$$\overline{X} = \frac{\displaystyle\sum_{i=1}^{n}(a + x_i)}{n} = \frac{\displaystyle\sum_{i=1}^{n} a}{n} + \frac{\displaystyle\sum_{i=1}^{n} x_i}{n}$$

$$= \frac{na}{n} + \overline{x} = a + \overline{x}$$

If S = the standard deviation of $X_1, X_2, X_3, \ldots, X_n$ and σ = the standard deviation of $x_1, x_2, x_3, \ldots, x_n$

$$S = \sqrt{\frac{\Sigma x^2}{n} - (\overline{X})^2}$$

$$= \sqrt{\frac{\Sigma (a + x)^2}{n} - \frac{\Sigma (a + x)^2}{n}}$$

$$= \sqrt{\frac{\Sigma\,(a^2 + 2ax + x^2)}{n} - \left(\frac{\Sigma\,a + \Sigma\,x}{n}\right)^2}$$

$$= \sqrt{\frac{\Sigma\,a^2 + \Sigma\,2ax + \Sigma\,x^2}{n} - \left(\frac{na}{n} + \frac{\Sigma\,x}{n}\right)^2}$$

$$= \sqrt{\frac{na^2}{n} + 2a\bar{x} + \frac{\Sigma\,x^2}{n} - (a + \bar{x})^2}$$

$$= \sqrt{a^2 + 2a\bar{x} + \frac{\Sigma\,x^2}{n} - a^2 - 2a\bar{x} - (\bar{x})^2}$$

$$= \sqrt{\frac{\Sigma\,x^2}{n} - (\bar{x})^2} = \sigma$$

Hence if a constant amount, a, is added to each number in a set of numbers the mean is increased by a but the standard deviation remains unaltered.

5. If the numbers $x_1, x_2, x_3, \ldots, x_n$ are all multiplied by the same amount, a, such that $X_1 = ax_1$, $X_2 = ax_2$, $X_3 = ax_3, \ldots, X_n = ax_n$, then

$$\bar{X} = \frac{\sum\limits_{i=1}^{n} ax_i}{n} = \frac{a \sum\limits_{i=1}^{n} x_i}{n} = a\bar{x}$$

If S = the standard deviation of $X_1, X_2, X_3, \ldots, X_n$ and σ = the standard deviation of $x_1, x_2, x_3, \ldots, x_n$

$$S = \sqrt{\frac{\Sigma\,(ax)^2}{n} - \left(\frac{ax}{n}\right)^2}$$

$$= \sqrt{\frac{\Sigma\,(ax)^2}{n} - \frac{ax}{n}^2}$$

$$= \sqrt{\frac{a^2\,\Sigma\,x^2}{n} - a^2(\bar{x})^2}$$

$$= a\sqrt{\frac{\Sigma\,x^2}{n} - (\bar{x})^2} = a\sigma$$

Hence if each number in a set of numbers is multiplied by a constant amount, a, both the mean and standard deviation of the original set of numbers is multiplied by a.

IMPORTANT FORMULAE AND DATA

ARITHMETIC MEAN

For a set of observations $x_1, x_2, \ldots, x_n$, the arithmetic mean is

$$\bar{x} = \frac{\Sigma x}{n}$$

If the observations occur with frequencies $f_1, f_2, \ldots, f_n$, the arithmetic mean is

$$\bar{x} = \frac{\Sigma fx}{n}$$

When an assumed mean is used,

$$\bar{x} = A + \frac{\Sigma fd}{n}$$

where A = assumed mean

and $d_i = x_i - A$

MEDIAN

For a series of observations arranged in descending (or ascending order) of magnitude, the median is the value of the middle observation. If the number of observations is even, the median is the arithmetic mean of the two middle observations.

For a frequency distribution, the median is obtained by drawing a cumulative frequency curve (an ogive). The median is then the value of the variable corresponding to half of the total frequency.

QUANTILES

The *quartiles* divide the data into four equal parts. The lower quartile, Q_1, is the value of the variable corresponding to one-quarter of the total frequency. The upper quartile, Q_3, is the value of the variable corresponding to three-quarters of the total frequency.

The *deciles* divide the data into ten equal parts. For instance, the sixth decile, D_6, is the value of the variable corresponding to six-tenths of the total frequency.

The *percentiles* divide the data into one-hundred equal parts. For instance, the 81st percentile is the value of the variable corresponding to eighty-one-hundredths of the total frequency.

MODE

For a set of observations the mode is the value of the most frequently occurring observation.

For a frequency distribution the mode may be found by drawing a histogram and performing the construction shown in Fig. 1.

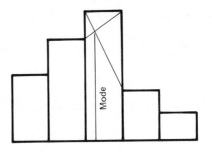

Fig. 1.

WEIGHTED ARITHMETIC MEAN

If a set of observations $x_1, x_2, \ldots, x_n$ have weights $w_1, w_2, \ldots, w_n$, the weighted arithmetic mean is

$$\bar{x} = \frac{\Sigma\, wx}{\Sigma\, w}$$

GEOMETRIC MEAN

The geometric mean of a set of observations, $x_1, x_2, \ldots, x_n$, is

$$G = \sqrt[n]{x_1 x_2 \ldots x_n}$$

For a frequency distribution,

$$G = \sqrt[n]{x_1^{f_1}\, x_2^{f_2} \ldots x_n^{f_n}}$$

where $n = f_1 + f_2 + \ldots + f_n$

SKEWED DISTRIBUTIONS

For moderately skewed distributions which are unimodal,

$$\text{mean} - \text{mode} = 3\,(\text{mean} - \text{median})$$

RANGE

$$\text{Range} = \text{largest observation} - \text{smallest observation}$$

MEAN DEVIATION

For a set of observations $x_1, x_2 \ldots, x_n$,

$$\text{mean deviation} = \frac{\Sigma \,|x - \bar{x}|}{n}$$

If the observations occur with frequencies $f_1, f_2, \ldots, f_n$,

$$\text{mean deviation} = \frac{\Sigma \, f|x - \bar{x}|}{n}$$

where $n = f_1 + f_2 + \ldots + f_n$

SEMI-INTERQUARTILE RANGE

Semi-interquartile range $= \frac{1}{2}(Q_3 - Q_1)$

where Q_1 = lower quartile

and Q_3 = upper quartile

STANDARD DEVIATION

For a set of observations $x_1, x_2, \ldots x_n$, the standard deviation is

$$\sigma = \sqrt{\frac{\Sigma \,(x - \bar{x})^2}{n}} = \sqrt{\frac{\Sigma \, x^2}{n} - (\bar{x})^2}$$

If the observations occur with frequencies $f_1, f_2, \ldots f_n$,

$$\sigma = \sqrt{\frac{\Sigma \, f(x - \bar{x})^2}{n}} = \sqrt{\frac{\Sigma \, fx^2}{n} - (\bar{x})^2}$$

LINE OF BEST FIT

The line of best fit approximating to the set of points $(x_1, y_1), (x_2, y_2) \ldots$ (x_n, y_n) passes through the point $(\bar{x}, \bar{y})$ where $\bar{x}$ is the mean of the x values and $\bar{y}$ is the mean of the y values.

The equation of a straight line is $y = a + bx$ where a is the intercept on the y-axis and b is the gradient of the line. b is called the regression coefficient.

The equation $y = a + b\sqrt{x}$ will give a straight line graph if y is plotted against $\sqrt{x}$.

The equation $y = a + \dfrac{b}{x}$ will give a straight line if y is plotted against $\dfrac{1}{x}$.

The equation $y = a + bx^2$ will give a straight line if y is plotted against x^2.

COEFFICIENT OF CORRELATION

The coefficient of correlation has a value between -1 and $+1$. For precise positive correlation the coefficient has a value of $+1$ and for precise negative correlation its value is -1. For uncorrelated data the value of the coefficient is zero.

If the equation of the regression line of y on x is $y = a + bx$ and the equation of the regression line of x on y is $x = a_1 + b_1 y$, then

$$\text{coefficient of correlation} = \sqrt{bb_1}$$

VARIANCE

$$\text{variance} = \sigma^2$$

where σ = the standard deviation

SKEWNESS

$$\text{Pearson's coefficient of skewness} = \frac{3(\text{mean} - \text{median})}{\text{standard deviation}}$$

The coefficient can take any value between -3 and $+3$. The higher the value the greater the degree of skewness. For a symmetrical distribution the value is zero.

RANK CORRELATION

Spearman's coefficient of rank correlation is given by

$$R = 1 - \frac{6\Sigma d^2}{n(n^2 - 1)}$$

where d = the difference in ranks of corresponding values of x and y

n = the number of pairs of values of x and y in the set

A First Course in Statistics

MOVING AVERAGES

For the set of numbers $x_1, x_2, x_3 \ldots$ the moving average of order n is given by the following sequence of arithmetic means:

$$\frac{x_1 + x_2 + \ldots + x_n}{n}, \quad \frac{x_2 + x_3 + \ldots + x_{n+1}}{n}, \quad \frac{x_3 + x_4 + \ldots + x_{n+2}}{n}$$

INDEX NUMBERS

$$\text{price relative} = \frac{p_n}{p_0} \qquad p_0 = \text{price at base period}$$

$$p_n = \text{price at given period;}$$

$$\text{quantity or volume relative} = \frac{q_n}{q_0}$$

$$q_0 = \text{quantity at base period}$$

$$q_n = \text{quantity at given period}$$

$$\text{value relative} = \frac{p_n q_n}{p_0 q_0}$$

PROBABILITY

All probabilities have values between 0 and 1, a probability of 0 representing an absolute impossibility and a probability of 1 representing an absolute certainty.

$$P(E) = \frac{\text{total number of favourable outcomes}}{\text{total number of possible outcomes}}$$

The total probability covering all possible events $= 1$

In cases where the probability has to be determined by experiment:

$$P(E) = \frac{\text{number of trials with favourable outcomes}}{\text{total number of trials}}$$

Independent events: Probability of *all* the events occurring is

$$P(E_1 E_2 \ldots E_n) = P(E_1) \times P(E_2) \times \ldots \times P(E_n)$$

Mutually exclusive events: Probability of *one* of the events occurring is

$$P(E_1 + E_2 + \ldots + E_n) = P(E_1) + P(E_2) + \ldots + P(E_n)$$

Non-mutually exclusive events: The probability that either E_1 or E_2 or both E_1 and E_2 will occur is

$$P = P(E_1) + P(E_2) - P(E_1) \times P(E_2) = 1 - P(\overline{E_1}) \times P(\overline{E_2})$$

where $P(\overline{E_1}) = 1 - P(E_1)$ and $P(\overline{E_2}) = 1 - P(E_2)$

ANSWERS

ANSWERS TO CHAPTER 1

EXERCISE 1

1. $x = 4; y = 4.8$
2. £100 000; £28 000; £20 000; £8 000
3. 15.7%; 39.3%; 45.0%
9. 38 p; 17 p; 12 p; 13 p; 20 p
10. 6.12 cm; 6.77 cm
13. 1 600 000 000; 2 500 000 000

ANSWERS TO CHAPTER 2

EXERCISE 2

10. (a) £1.71   (b) £3.17
11. (a) 8   (b) 3.7

MISCELLANEOUS EXERCISE

EXERCISE 3

2. 27 years; 11 years; 63 years
5. 17; 4.5 and 7.5
6. (a) $132°; 36°; 24°, 131°, 37°$
7. $168°, 72°, 96°, 24°$
8. (a) $315°; 35°; 10°$
   (c) 5.14 cm
10. c 11. b 12. c 13. c 14. c 15. c

ANSWERS TO CHAPTER 3

EXERCISE 4

1. (a) (i) 24.87   (ii) 25
   (iii) 24.8658
   (b) (i) 0.0084   (ii) 0.0083
   (iii) 0.008 36
   (c) (i) 4.9785   (ii) 4.98   (iii) 5
   (d) 22   (e) 35.60
   (f) (i) 28 388 000   (ii) 28 000 000
   (g) (i) 4.1497   (ii) 4.150   (iii) 4.15
   (h) (i) 9.205   (ii) 9.20

2. (a) (i) 5.1499   (ii) 5.150   (iii) 5.15
   (b) (i) 35.29   (ii) 35.286   (iii) 35.3

(c) (i) 0.004 98   (ii) 0.0050
   (iii) 0.005
(d) (i) 8.408   (ii) 8.41   (iii) 8.4
(e) (i) 0.853   (ii) 0.85

EXERCISE 5

1. 1199; 1201   2. 19 910; 19 950
3. 17.635; 17.665   4. 6; 8
5. 24 000; 26 000   6. 87; 89
7. 393.75; 434.75   8. 18.83; 18.95
9. 267 750 000; 304 750 000
10. 4.69; 4.73   11. 0.22; 0.24
12. 1.37; 1.39   13. 0.05 cm
14. 0.5 s; 0.14%   15. 4.38%
16. 4.4; 3.5 17. (a) 5   (b) 0.75
18. (a) 1.1   (b) 8.47   (c) 0.16
    (d) 3.297   (e) 22   (f) 30
    (g) 0.162   (h) 27.2
19. (8.56, 8.11 m/s)

ANSWERS TO CHAPTER 4

EXERCISE 6

1.

Mark	Frequency
1	2
2	2
3	11
4	11
5	12
6	7
7	4
8	1

2.

Age	Frequency
13	2
14	10
15	6
16	6
17	4
18	2

3.

Time(s)	Frequency
24	1
25	2
26	4
27	3
28	0
29	2
30	1
31	3
32	4

4.

Score	Frequency
0 − 3	4
4 − 7	8
8 − 11	19
12 − 15	10
16 − 19	5
20 − 23	2
24 − 27	1
28 − 31	1

5.

Mark	Frequency
0 − 9	0
10 − 19	3
20 − 29	7
30 − 39	10
40 − 49	16
50 − 59	34
60 − 69	13
70 − 79	7
80 − 89	6
90 − 99	4

7.

Height	Frequency
150	2
151	3
152	4
153	5
154	6
155	7
156	8
157	6
158	5
159	4
160	3
161	1

9. (a) continuous   (b) continuous
   (c) discrete   (d) discrete
10. 64
11. (a) 14.985; 15.015 mm
    (b) 0.03 mm
12. (a) continuous   (b) discrete
    (c) continuous   (d) discrete
    (e) continuous
16. 50; 60; 40;

EXERCISE 7

4. (a)

Pocket money (pence)	Number of children
1 − 9	0
10 − 19	5
20 − 29	7
30 − 39	17
40 − 49	26
50 − 59	16
60 − 69	10
70 − 79	10
80 − 89	5
90 − 99	4

(b)

Pocket money (p) less than	Number of children
10	0
20	5
30	12
40	29
50	55
60	71
70	81
80	91
90	96
100	100

(c) 60 children
5. frequency curve

6.

Noise level (dBA)	Frequency
81 − 85	4
86 − 90	12
91 − 95	9
96 − 100	7
101 − 105	4

Less than	No. of discos
81	0
86	4
91	16
96	25
101	32
106	36

42% of discos had a noise level exceeding 94 dBA

7. (a)

Less than	Cum. frequency
10	5
20	38
30	120
40	268
50	538
60	753
70	840
80	884
90	896
100	900

(b) (i) 20 (ii) 100
(c) 11.1

10.

Less than £	Cum. frequency
20	3
30	22
40	50
50	72
60	85
70	91
80	95
90	97

(a) 18% (b) 36% (c) 4%

11. 48%

ANSWERS TO CHAPTER 5

EXERCISE 8

1. £28 2. 174 cm 3. £82.40 4. 3.8
5. 2 6. 11 7. 199.92 mm
8. 9.85 9. 34.9
10. 19°C

EXERCISE 9

1. £1.00 per kg 2. 47 3. 4 4. 18.58
5. £0.97 per kg 6. 1.2487 7. 162.97

EXERCISE 10

1. 5 2. 4.5 3. 57 4. £4648
6. (c)(i) 85.2 (ii) 80.0 (iii) 91.0
7. 18; 26 8. 1 call per day
9. (a) 52 (b) 76 (c) 30 (d) 25th
10. (b) 138; 142; 134 cm
11. (a) 3½ kg (b) 2½ kg (c) 2½ kg
    (d) 1¾ kg (e) 3½ kg
12. (a) (i) 100 hours (ii) 701 −800 hours
    (b) 20.5% (d) 717 hours

13. (a)

Maximum temperature (°C)	Frequency
2	3
3	5
4	3
5	5
6	3
7	6

(c) (i) 7°C (d) 5°C
14. 6
15. (a)

Hand spar	Frequency
13	1
14	0
15	2
16	4
17	7
18	9
19	8
20	4
21	1
22	2
23	1
24	0
25	1

(c) 18 cm
16. (a) 53 (b) 48 (c) 175 (d) 55
    (e) 93%
18. 7 19. median
20. (a) 2 (b) 2

ANSWERS TO CHAPTER 6

EXERCISE 11

1. £42.80   2. 29.76 cm   3. 6   4. 2
5. 0.49   6. 0.706
7. (a)

Age group	Frequency
Under 10	9
Under 20	17
Under 30	24
Under 40	31
Under 50	38
Under 60	45
Under 70	50
Under 80	53

   (c) 16 years, 33 years, 53 years
   (d) 18.5 years
8. (a) 2½ kg   (b) 2 kg   (c) 1½ kg
   (d) 1½ kg   (e) 2½ kg   (f) 0.5 kg
9. (b) 310; 350   (c) 20
10. 4; 2.160
11. 13; 3.162
   (a) 130; 31.62   (b) 150; 31.62
12. 17.63 cm; 1.121 cm
13. 6.014 mm; 0.0276 mm
14. (a) 168.2 cm; 6.187 cm
   (b) 168.7 cm; 6.197 cm
15. (a) 21 and 6   (b) 12 and 2
16. 0.3   17. In English
18. (a) 127   (b) 65
19. 64; 62; 78   20. 88
21. (a) 9   (b) 36   22. 1.5
23. − 3   24. (b) 1.5
25. (a) 3 kg   (b) 55 kg   (c) 63 kg
   (d) 50%
26. (a) 8 cm;   (c) 38.5 cm; 1.2 cm
27. (a) 20, 32, 70   (b) 40, 51, 65
   (c) (i) 25 (ii) 12.5   (d) 87
28. (a) 10   (b) 8   (c) 2.828
29. (a) 39, 41, 43, 46   (b) 41.0 g, 2.21 g
30. (a) 13 hours   (b) 16 hours
   (c) 11 hours   (d) 16 hours
   (e) 2.5 hours

MISCELLANEOUS EXERCISE

EXERCISE 12

1. 98 m/s and 102 m/s
2. 160 cm and 190 cm
3. 69 kg   4. (a) 4   (b) 16
5. 11 − 15   9. 8.9 and 11.5

12. 11   13. 9%
16. 7
17. $\dfrac{0- \quad 5- \quad 15-35}{500 \quad 400 \quad 200}$
19. 0.75
20. (a) 31.5 − 33.5 cm
   (b) 31.95 − 33.95 mm
21. (a) 16   (b) 24   22. 52%
23. 1   24. 11
26. 1   27. 3.22
28. 28 years   29. (a) 14   (b) 4
30. 48   33. 164
34. 12.3 years
37. (a) 63 (b) 5.5 (c) 67 (d) 96 (e) 82
38. 47.3, 18; 52.5, 19.4
39. (b) 277 mm; 3 mm   (c) 33
40. (a) 112   (b) 105   41. 108 m; 112 m
42. (a)

Age group	Frequency
Under 10	18
Under 20	34
Under 30	48
Under 40	62
Under 50	76
Under 60	90
Under 70	100
Under 80	106

   (c) (i) 33 years (ii) 53 years
       (iii) 16 years (iv) 18.5 years
   (d) 28 millions
43. 34.30 years; 23.2 years
44. 224 g; 231 g; 227 g; 3.5 g
45. 31.4; 33.4; 9.4; 0.64
46. 1970: median   = £11.500
         mean   = £12.500
         standard = £ 4.000
         skewness =   0.75
    1980: median   = £14.000
         mean   = £16.000
         standard = £ 5.700
         skewness =   0.35
47. (b) 6   (c) 7
48. (a) 89.5 mm; 109.5 mm
   (c) 99 mm; 1.2 mm   (d) 4%
49. (a) £78 (i) 45 (ii) 15 (iii) 5
   (b) semi-interquartile range = £10.5
51. d   52. c   53. b   54. d   55. b
56. c   57. b   58. c   59. b   60. d
61. c   62. b   63. b   64. c   65. b
66. d   67. d   68. b   69. c   70. d
71. a   72. b   73. d   74. b   75. a
76. a   77. b   78. d

ANSWERS TO CHAPTER 7

EXERCISE 13

2. (a) 1; 3   (b) 3; 4   (c) − 5; 12
3. $a = 7$; $b = 4$
4. $a = 1$; $b = 2$
5. $a = - 2$; $b = 3$
6. $a = 4$; $b = 3$
7. $a = 0.8$; $b = 4.8$
8. $a = 10$; $b = 20$; £90
9. $a = 17.7$; $b = 0.77$
10. $a = - 1.2$ $b = 2.1$
11. $a = 5.3$; $b = 4.4$
12. $a = 0.5$; $b = 1.5$
13. $a = 248$; $b = 406$;   £262
14. $k = 98$; $C = 4.4$ $B = 19$

ANSWERS TO CHAPTER 8

EXERCISE 14

1. 0.9165; excellent   2. (b)
3. (a) − 0.6   (b) 0.8   (c) 0.002
4. 0   6. 1   7. 1   8. (a) − 7 (b) 76
9. $y = 5.0$
10. (b) 51   (c) 57.7   (f) $f = 19.4 + 0.75 g$
11. 5.5
12. $B = 7.49$; $D = 4.16$;
    $G = -4.16$; $I = -8.64$
    (b) $y = 82 - 0.49x$ (c) − 0.9030
13. 0.2857   14. 0.2
15. (a) 0.9152   (b) 0.7576
    (c) applicant A
16. (b) 44.17; 52.5   (d) 55; 83

ANSWERS TO CHAPTER 9

EXERCISE 15

1. 16 per thousand   2. 1680   3. 79
4. 13.56; 13.43
5. (a) 87.5   (b) 89.0   (c) 96.5
6. 72; 75.1; 90.5; 89.4

ANSWERS TO CHAPTER 10

EXERCISE

1. (a) 133.3   (b) 75

2. (a)

1975	1976	1977	1978	1979	1980
100	117	140	167	200	240

(b)

1975	1976	1977	1978	1979	1980
60	70	84	100	120	144

3.

1976	1977	1978	1979	1980
100	102	106	111	119

4. (a) 160   (b) 62.5
5.

1955	1960	1965	1970
100	263	355	428

6. 133   7. 139.7
8. 94%   9. 164
10. 61.9   11. 128
12. 1974: 93.9   1975: 95.0
13. 103.8
14.

1974	1975	1976	1977	1978	1979
100	104	110	119	113	116

15. 105; 110   16. 109; 121
17.

1976	1977	1978	1979	1980
100	110	127	149	182

ANSWERS TO CHAPTER 11

EXERCISE 17

1. 4; 5; 3.5; 2; 1.5
2. 4; $4^{1}/_{3}$; $2^{2}/_{3}$; 2
4. $y = 44.5 - 0.77x$; 35.3
5. $y = 54.1 + 13.2x$; £265 300
6.

Quarters	Seasonal factors	Year 2
1	88.1	936.9
2	− 70.1	950.1
3	− 52.3	862.3
4	34.3	875.7

7.

Quarters	1	2	3	4
Seasonal variation	19	− 17	− 19	19

(c) (43, 78)

ANSWERS TO CHAPTER 12

EXERCISE 18

1. (a) $\dfrac{1}{6}$ (b) $\dfrac{1}{3}$ (c) $\dfrac{1}{2}$

2. (a) $\dfrac{1}{52}$ (b) $\dfrac{1}{13}$ (c) $\dfrac{4}{13}$

   (d) $\dfrac{1}{26}$

3. (a) $\dfrac{1}{4}$ (b) $\dfrac{3}{8}$ (c) $\dfrac{5}{8}$

4. (a) $\dfrac{1}{5}$ (b) $\dfrac{1}{2}$ (c) $\dfrac{7}{10}$

5. (a) $\dfrac{1}{9}$ (b) $\dfrac{1}{6}$ (c) $\dfrac{13}{18}$

6. (a) $\dfrac{1}{15}$ (b) $\dfrac{1}{15}$ (c) $\dfrac{1}{10}$

7. (a) $\dfrac{7}{50}$ (b) $\dfrac{9}{25}$ (c) $\dfrac{1}{10}$

8. (a) $\dfrac{7}{50}$ (b) $\dfrac{6}{50}$ (c) $\dfrac{7}{10}$

EXERCISE 19

1. $\dfrac{5}{52}$ 2. (a) $\dfrac{1}{12}$ (b) $\dfrac{1}{4}$

3. (a) $\dfrac{4}{25}$ (b) $\dfrac{9}{25}$ (c) $\dfrac{12}{25}$

4. $\dfrac{3}{100}$ 5. (a) 0.32   (b) 0.0273

6. (a) $\dfrac{1}{13}$ (b) $\dfrac{1}{17}$

7. (a) $\dfrac{4}{27}$ (b) $\dfrac{5}{27}$

8. (a) $\dfrac{1}{3}$ (b) $\dfrac{5}{12}$

   (c) (i) $\dfrac{1}{24}$ (ii) $\dfrac{1}{12}$

9. (a) $\dfrac{1}{6}$ (b) (i) $\dfrac{1}{36}$ (ii) $\dfrac{1}{18}$ (c) $\dfrac{1}{6}$

10. (a) $\dfrac{1}{6}$ (b) $\dfrac{1}{2}$ (d) (i) $\dfrac{1}{9}$ (ii) $\dfrac{5}{18}$

11. (a) $\dfrac{1}{10}$ (b) $\dfrac{3}{10}$ (c) $\dfrac{7}{10}$

   (d) $\dfrac{1}{15}$ (e) $\dfrac{8}{15}$

12. (a) $A = \dfrac{5}{12}$; $B = \dfrac{5}{11}$; $C = \dfrac{7}{11}$;

   $D = \dfrac{4}{11}$;   (b) $\dfrac{7}{22}$

13. (a) $\dfrac{4}{35}$ (b) $\dfrac{4}{35}$ (c) $\dfrac{6}{35}$

   (d) $\dfrac{12}{35}$ (e) $\dfrac{4}{35}$

14. (b) $0; \dfrac{1}{25}; \dfrac{2}{25}; \dfrac{1}{5}; \dfrac{1}{25}$

   (c) (i) 4   (ii) 9   (iii) 16

15. (a) $q^5 + 5q^4p + 10q^3p^2 + 10q^2p^3 + 5qp^4 + p^5$

   (b) $q^7 + 7q^6p + 21q^5p^2 + 35q^4p^3 + 35q^3p^4 + 21q^2p^5 + 7qp^6 + p^7$

   (c) $q^9 + 9q^8p + 36q^7p^2 + 84q^6p^3 + 126q^5p^4 + 126q^4p^5 + 84q^3p^6 + 36q^2p^7 + 9qp^8 + p^9$

16. (a) $\dfrac{1}{16}$ (b) $\dfrac{4}{16}$ (c) $\dfrac{6}{16}$

   (d) $\dfrac{4}{16}$ (e) $\dfrac{1}{16}$

17. (a) $\dfrac{1}{169}$ (b) $\dfrac{9}{169}$

18. (a) 0.8145   (b) 0.0135   (c) 0.0140

MISCELLANEOUS EXERCISE

EXERCISE 22

1. 135   2. 1

3.

1976	1977	1978	1979
100	120	125	110

4. $\dfrac{2}{7}$ 5. 0.5   6. 115   8. £7000

9. 0 and 1   10. 4.5   11. 15

12. $y = 2x − 11$ 13. 354   15. 25.25

16. (a) 90   (b) 2.5   (c) $\dfrac{11}{18}$

17. (a) 0.9429   (c) 21.0, 34.0

   (d) $y = 1.21x + 10$ (e) 33.0

18. (d) £9350   19. (c) 50

21. (a) (i) $\dfrac{1}{3}$ (ii) $\dfrac{2}{3}$ (iii) $\dfrac{2}{3}$

   (iv) $\dfrac{1}{6}$ (b) $\dfrac{1}{5}$

22. (a) $A = \dfrac{10}{24}$; $B = \dfrac{10}{23}$; $C = \dfrac{14}{23}$

   $D = \dfrac{9}{23}$ (b) $\dfrac{15}{92}$

26. (a) 1.6; − 9.9   (b) 0.77

27. (b) $\frac{2}{36} = \frac{1}{18}$ (c) $\frac{6}{36} = \frac{1}{6}$ (d) $\frac{25}{36}$

    (e) £2.60

28. (a) 3214   (b) 4625

29. 15.9; 21.1; 11.4; 21.2

30. $y = 2.97\sqrt{x} + 2.10$; 28.8

31. (a) (i) $\frac{4}{9}$ (ii) $\frac{1}{3}$ (iii) $\frac{2}{9}$

    (b) $\frac{29}{81}$; $\frac{52}{81}$ (c) 12

32.

% change	Weights	Weight x (%)
25	4	100
20	3	60
10	2	20

Cost in Rins	Suitable weights
80	$A = 4$
60	$B = 3$
40	$C = 2$

    (c) 120

33. (b) 54; 59   (d) 44

34. *b* 35. *b* 36. *c* 37. *c* 38. *c* 39. *a*

40. *d* 41. *d* 42. *d* 43. *c* 44. *c* 45. *d*

46. *b* 47. *c* 48. *c* 49. *c* 50. *a*

INDEX